Gas-Liquid and Liquid-Liquid Separators

Gas-Liquid and Liquid-Liquid Separators

Maurice Stewart
Ken Arnold

ELSEVIER

AMSTERDAM • BOSTON • HEIDELBERG • LONDON • NEW YORK • OXFORD
PARIS • SAN DIEGO • SAN FRANCISCO • SINGAPORE • SYDNEY • TOKYO

Gulf Professional Publishing is an imprint of Elsevier

Gulf Professional Publishing is an imprint of Elsevier
30 Corporate Drive, Suite 400, Burlington, MA 01803, USA
Linacre House, Jordan Hill, Oxford OX2 8DP, UK

Library of Congress Cataloging-in-Publication Data
Application submitted

British Library Cataloguing-in-Publication Data
A catalogue record for this book is available from the British Library.

ISBN: 978-0-7506-8979-3

For information on all Gulf Professional Publishing publications visit our Web site at www.elsevierdirect.com

Printed in the United States of America
08 09 10 10 9 8 7 6 5 4 3 2 1

Contents

A Note from the Authors

Gulf Equipment Guides series serves as a quick reference for the design, selection, specification, installation, operation, testing, and trouble-shooting of surface production equipment. The *Gulf Equipment Guides* series consists of multiple volumes, each of which covers a specific area in surface production equipment. These guides cover essentially the same topics included in the "Surface Production Operations" series but omit the proofs of equations, example problems and solutions which belong more properly in a handbook. This book contains fewer pages and is therefore more focused. The reader is referred to the corresponding volume of the "Surface Production Operations" series for further details and additional information such as derivations of some of the equations, example problems and solutions and suggested test questions.

About the Book

Gas–Liquid and Liquid–Liquid Separators is the first volume in the *Gulf Equipment Guides* series. Each guide serves as a quick reference resource. The series is intended to provide the most comprehensive coverage you'll find today dealing with surface production operation in its various stages, from initial entry into the flowline through gas–liquid and liquid–liquid separation; emulsions, oil and water treating; water injection; hydrate prediction and prevention; gas dehydration; and gas conditioning and processing equipment to the exiting pipeline.

Featured in this volume are such important topics as basic principles, process selection, gas–liquid separators, liquid–liquid separators, and mechanical design of pressure vessels. This volume as well as all volumes in the *Gulf Equipment Guides* series, serve the practicing engineer and senior field personnel by providing organized design procedures; details on suitable equipment for application selection; and charts, tables, and nomographs in readily useable form. Facility engineers, process engineers, designers, operations engineers, and senior production operators will develop a "feel" for the important parameters in designing, selecting, specifying, and trouble-shooting surface production facilities. Readers will understand the uncertainties and assumptions inherent in designing and operating the equipment in these systems and the limitations, advantages, and disadvantages associated with their use.

CHAPTER 1

Basic Principles

1.1 Introduction

Before describing gas–liquid (2-phase) and liquid–liquid (3-phase) separation equipment used in oil and gas production facilities and design techniques for selecting and sizing that equipment, it is necessary to review some basic principles and fluid properties. We will also discuss some of the common calculation procedures, conversions, and operations used to describe the fluids encountered in the production operations.

1.2 Fluid Analysis

An example fluid analysis of a typical gas well is shown in Table 1.1. Note that only paraffin hydrocarbons are shown. This is not correct, even though they may be the predominant series present. Also note that all molecules of heptane and larger ones are lumped together as heptanes plus fraction.

1.3 Physical Properties

An accurate estimate of physical properties is essential if one is to obtain reliable calculations. Physical and chemical properties depend upon:

- Pressure
- Temperature
- Composition

Most hydrocarbon streams are mixtures of hydrocarbons that may contain varying quantities of contaminants such as

- Hydrogen sulfide
- Carbon-dioxide
- Water

TABLE 1.1
Example fluid analysis of gas well

Component	*mol %*
Methane (C_1)	35.78
Ethane (C_2)	21.46
Propane (C_3)	1.40
i-Butane (*i*-C_4)	5.35
n-Butane (*n*-C_4)	10.71
i-Pentane (*i*-C_5)	3.81
n-Pentane (*n*-C_5)	3.07
Hexanes (C_6)	3.32
Heptanes plus (C_{7+})	3.24
Nitrogen	0.20
Carbon dioxide	1.66
Total	**100.00**

The more similar the character of the mixture molecules, the more orderly their behavior. A single component system composed entirely of a simple molecule, like methane, behaves in a very predictable, correctable manner.

The accuracy of calculations decrease in the following order:

- Single component system
- Mixture of molecules from the same homologous series
- Mixture of molecules from different homologous series
- Hydrocarbon mixtures containing sulfur compounds and/or carbon dioxide

Performance data for a single component system can be accurately correlated in graphical or tabular form. For all others, one must use either pressure/volume/temperature (PVT) equations of state or the Weighted Average. The Weighted Average assumes that the contribution of an individual molecule is in proportion to its relative quantity in the mixture. The more dissimilar the molecules, the less accurate the prediction becomes. Table 1.2 lists some of the physical properties of some of the paraffin hydrocarbon series.

Water in liquid or vapor form is present to some degree in all systems. Liquid water is essentially immiscible in hydrocarbons. However, in the vapor phase it represents a small percentage (seldom more than one part per thousand, by weight). Since normal phase behavior calculations do not apply for water, special procedures must be used. Equations of state use the values of P, V, and T at the critical point. Each molecular species has a unique critical point.

TABLE 1.2
Physical properties of paraffin hydrocarbons

Component	*Methane*	*Ethane*	*Propane*	*iso-Butane*	n-*Butane*	*iso-Pentane*	n-*Pentane*	n-*Hexane*	n-*Heptane*	n-*Octane*	n-*Nonane*	n-*Decane*
Molecular weight	16.043	30.070	44.097	58.124	58.124	72.151	72.151	86.178	100.205	114.232	128.259	142.286
Boiling point @ 14.696 psia, °F	−258.73	−127.49	−43.75	10.78	31.08	82.12	96.92	155.72	209.16	258.21	303.47	345.48
Freezing point @ 14.696 psia, °F	−296.44	−297.49	−305.73	−255.28	−217.05	−255.82	−201.51	−139.58	−131.05	−70.18	−64.28	−21.36
Vapor pressure @ 100°F, psia	(5000.)	(800.)	188.4	72.58	51.71	20.445	15.574	4.960	1.620	0.5369	0.1795	0.0609
Density of liquid @ 60°F and 14.696 psia												
Relative density @ 60°F/60°F	(0.3)	0.3562	0.5070	0.5629	0.5840	0.6247	0.6311	0.6638	0.6882	0.7070	0.7219	0.7342
°API	(340.)	265.6	147.3	119.8	110.7	95.1	92.7	81.60	74.08	68.64	64.51	61.23
Absolute density, lbm/gal (in vacuum)	(2.5)	2.970	4.227	4.693	4.870	5.208	5.262	5.534	5.738	5.894	6.018	6.121
Apparent density, lbm/gal (in air)	(2.5)	2.960	4.217	4.683	4.861	5.198	5.252	5.524	5.729	5.885	6.008	6.112
Density of gas @ 60°F and 14.696 psia												
Relative density (air = 1), ideal gas	0.5539	1.0382	1.5225	2.0068	2.0068	2.4911	2.4911	2.9755	3.4598	3.9441	4.4284	4.9127
lb/M ft^3, ideal gas	42.28	79.24	116.20	153.16	153.16	190.13	190.13	227.09	264.06	301.02	337.98	374.95
Volume @ 60°F and 14.696 psia												
Liquid, gal/lb-mol	(6.4)	10.13	10.43	12.39	11.94	13.85	13.72	15.57	17.46	19.38	21.31	23.45
Ft^3 has/gal liquid, ideal gas	(59.1)	37.48	36.375	30.64	31.79	27.39	27.67	24.37	21.73	19.58	17.81	16.33

(Continued)

TABLE 1.2 (*Continued*)

Component	*Methane*	*Ethane*	*Propane*	*iso-Butane*	n-*Butane*	*iso-Pentane*	n-*Pentane*	n-*Hexane*	n-*Heptane*	n-*Octane*	n-*Nonane*	n-*Decane*
Ratio, gas/liquid, in vacuum	(442.)	280.4	272.1	229.2	237.8	204.9	207.0	182.3	162.6	146.5	133.2	122.2
Critical conditions												
Temperature, °F	−116.67	89.92	206.06	274.46	305.62	369.10	385.8	453.6	512.7	564.22	610.68	652.0
Pressure, psia	666.4	706.5	616.0	527.9	550.6	490.4	488.6	436.9	396.8	360.7	331.8	305.2
Gross calorific value, combustion @ 60°F												
Btu/lb, liquid	–	22181	21489	21079	21136	20891	20923	20783	20679	20607	20543	20494
Btu/lb, gas	23891	22332	21653	21231	21299	21043	21085	20942	20838	20759	20700	20651
Btu/ft^3, ideal gas	1016.0	1769.6	2516.1	3251.9	3262.3	4000.9	4008.9	4755.9	5502.5	6248.9	6996.5	7742.9
Btu/gal, liquid	–	65869	90830	98917	102911	108805	110091	115021	118648	121422	123634	125448
Volume air to burn one volume, ideal gas	9.54	16.71	23.87	31.03	31.03	38.19	38.19	45.35	52.52	59.68	66.84	74.00
Flammability limits @ 100°F and 14.696 psia												
Lower, volume % in air	5.0	2.9	2.0	1.8	1.5	1.3	1.4	1.1	1.0	0.8	0.7	0.7
Upper, volume % in air	15.0	13.0	9.5	8.5	9.0	8.0	8.3	1.7	7.0	6.5	5.6	5.4
Heat of Vaporation @ 14.696 psia												
Btu/lb @ boiling point	219.45	211.14	183.01	157.23	165.93	147.12	153.57	143.94	163.00	129.52	124.36	119.65
Specific heat @ 60°F and 14.696 psia												
C_p gas, Btu/(lb-°F), ideal gas	0.5267	0.4078	0.3885	0.3867	0.3950	0.3844	0.3882	0.3863	0.3845	0.3833	0.3825	0.3818
C_v gas, Btu/(lb-°F), ideal gas	0.4029	0.3418	0.3435	0.3525	0.3608	0.3869	0.3607	0.3633	0.3647	0.3659	0.3670	0.3678
$K = C_p/C_v$, ideal gas	1.307	1.193	1.131	1.097	1.095	1.077	1.076	1.064	1.054	1.048	1.042	1.038
C_p liquid, Btu/(lb-°F)	–	0.9723	0.6200	0.5707	0.5727	0.5333	0.5436	0.5333	0.5280	0.5241	0.5224	0.5210

For each of the pure components shown in the tables, the critical values represent the maximum pressure and temperature at which a two-phase, vapor–liquid system can exist. Above P_c and T_c, only a single phase is possible. For mixtures, pseudo-critical values are calculated, which are correlation constants only and are not a point on the phase diagram.

1.3.1 Equations of State

The correlations that follow are commonly used for hydrocarbon systems and are suitable for use for most calculations. Any equation correlating P, V, and T is called an equation of state. The ideal equation of state is sometimes called ideal gas law, perfect gas law, or general gas law and is expressed by Equation (1.1).

$$PV = nRT \tag{1.1}$$

where

P = absolute pressure
V = volume
n = number of moles of gas of volume V at P and T
R = Universal gas constant (refer to Table 1.3)
T = absolute temperature

Equation (1.1) is valid up to pressures of about 60 psia (500 kPa, 4 bara). As pressure increases above this level, its accuracy becomes less and the system should be considered a non-ideal gas. Table 1.3 lists the values of the universal gas constant for different unit systems.

1.3.2 Molecular Weight and Apparent Molecular Weight

The number of moles is defined as follows:

$$\text{Mole} = \frac{\text{Mass}}{\text{Molecular weight}} \tag{1.2}$$

TABLE 1.3
Universal gas constant

P	V	T	R
kPa	m^3	K	8.314 (kPa)(m^3)/(kmol)(K)
MPa	m^3	K	0.00831 (MPa)(m^3)/(kmol)(K)
bar	m^3	K	0.08314 (bar)(m^3)/(kmol)(K)
psi	ft^3	°R	10.73 (psia)(ft^3)/(lb–mol)(°R)
lb/ft^2	ft^3	°R	1545 (psia)($ft^3$1/(lb–mol)(°R)

expressed as

$$n = \frac{m}{M} \tag{1.3}$$

or in units as

$$\text{lb} - \text{mole} = \frac{\text{lb}}{\dfrac{\text{lb}}{\text{lb} - \text{mole}}} \tag{1.4}$$

Molecular weight is defined as the sum of the atomic weights of the various elements present.

Example 1.1: Molecular Weight Calculation

Given:

Determine the molecular weight of ethane, C_2H_6

Solution:

Element	*No. of Atoms*		*Atomic Weight*		*Product*
C	2	×	12	=	24
H	6	×	1	=	6
Molecular weight				=	30 lb/(lb–mol)

Up to now, we have addressed only pure substances. We now have to consider hydrocarbon mixtures. However, first we must discuss apparent molecular weight and specific gravity. It is not correct to say that a hydrocarbon mixture has a molecular weight; rather, it is an apparent molecular weight. Apparent molecular weight is defined as the sum of the products of the mole fractions of each component times the molecular weight of that component. This is shown in Equation (1.5):

$$\text{MW} = \sum y_i(\text{MW})_i \tag{1.5}$$

where

y_i = molecular fraction of *i*th component
MW_i = molecular weight of *i*th component
$\sum y_i = 1$

Now, let us look at an example of the application of apparent molecular weight that will also result with a number that we will use often throughout this book.

Example 1.2: Determine the apparent molecular weight of dry air, which is a gas mixture consisting of nitrogen, oxygen, and small amounts of Argon

Given:

Determine he apparent molecular weight of air given its approximate composition

Gas Composition	
Component	Mole fraction
Nitrogen	0.78
Oxygen	0.21
Argon	0.01
Total	1.00

Solution:

1. Look up the molecular weight of each component from the physical constant table

$$(MW)_N = 28; \quad (MW)_O = 32; \quad (MW)_A = 40$$

2. Multiply the mole fraction of each component by its molecular weight

$$(MW)_{AIR} = \sum y_i (MW)_i = y_N (MW)_N + y_O (MW)_O + y_A (MW)_A$$

$$= (0.78 \times 28) + (0.21 \times 32) + (0.01 \times 40) = 29 \text{ lb}/(\text{lb} - \text{mole})$$

We will now define the specific gravity of a gas.

1.3.3 Gas Specific Gravity

The specific gravity of a gas is the ratio of the density of the gas to the density of air standard conditions of temperature and pressure.

$$S = \frac{\rho_g}{\rho_{air}} \tag{1.6}$$

where

ρ_g = density of gas
ρ_{air} = density of air

Both densities must be computed at the same pressure and temperature, usually at standard conditions.

It may be related to the molecular weight by Equation (1.7).

$$S = \frac{(MW)_g}{29} \tag{1.7}$$

Example 1.3: Calculate the specific gravity of a natural gas with the following composition

Given:

Component	*Mole Fraction* (y_i)
Methane (C_1)	0.85
Ethane (C_2)	0.09
Propane (C_3)	0.04
n-Butane (*n*-C_4)	0.02
	1.00

Solution:

(1)

Component	*Mole Fraction*, y_i		*Molecular Weight*, $(MW)_i$		y_i $(MW)_i$
C_1	0.85	×	16.0	=	13.60
C_2	0.09	×	30.1	=	2.71
C_3	0.04	×	44.1	=	1.76
n-C_4	0.02	×	58.1	=	1.16
	1.00		$(MW)_g$	=	19.23

(2)

$$S = \frac{(MW)_g}{29} = \frac{19.23}{29} = 0.66$$

1.3.4 Non-Ideal Gas Equations of State

The ideal gas equations of state describe most real gases at low pressure but do not yield reasonable results at higher pressures. Many PVT equations have been developed to describe non-ideal, real gas behavior. Each is empirical in that it correlates a specific set of data using one, or more, empirical constants. Unfortunately, there is no correlation that is equally good for all gas mixtures. There can be as many such equations as there are individuals who correlate data. In some instances, the equations have been extrapolated beyond the compositions on which they were determined. This results in an inherent loss of accuracy.

The ideal equations of state can be approximated to the compressibility equation of state by multiplying the "RT" part of the equation by Z:

$$PV = ZnRT \tag{1.8}$$

where

$$Z = \frac{\text{Actual gas volume}}{\text{Ideal gas volume}} \tag{1.9}$$

If the gas acted as if it were an ideal gas, then the "Z" factor would be 1. The typical range of $Z = 0.8$–1.2.

The compressibility factor for a natural gas can be approximated from Figures 1.1 through 1.6, which are from the *Engineering Data Book* of the Gas Processor Suppliers Association.

1.3.5 Liquid Density and Specific Gravity

The specific gravity of a liquid is the ratio of the density of the liquid at 60 °F to the density of pure water.

$$\text{SG} = \frac{\rho_1}{\rho_\text{w}} \tag{1.10}$$

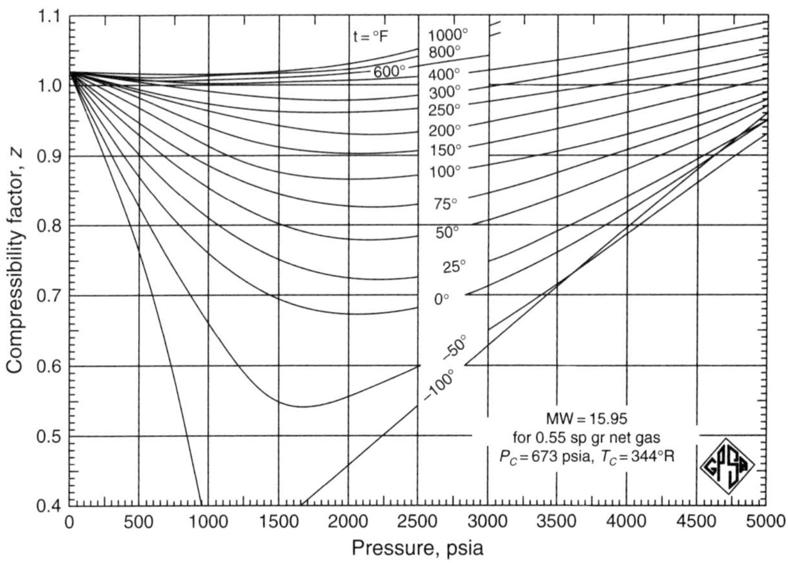

FIGURE 1.1. Compressibility factor for specific gravity = 0.55 gases (courtesy of GPSA engineering Data Book).

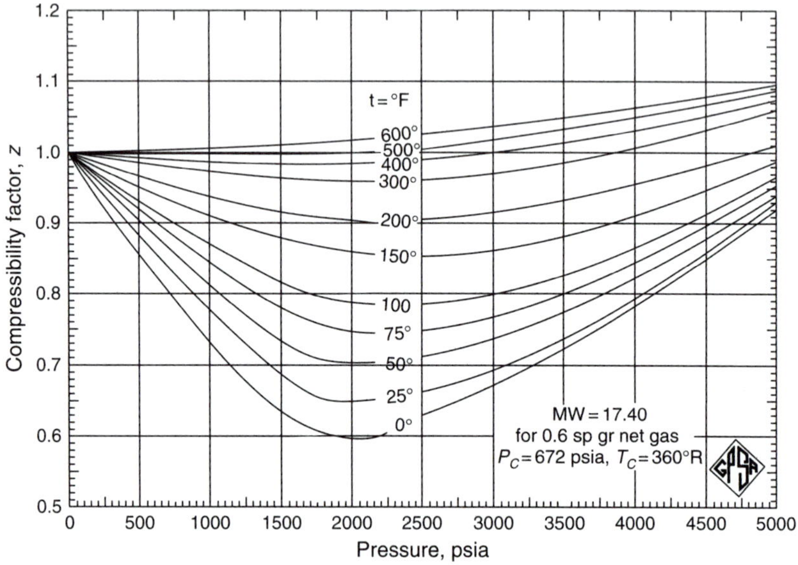

FIGURE 1.2. Compressibility factor for specific gravity = 0.6 gases (courtesy of GPSA Engineering Data Book).

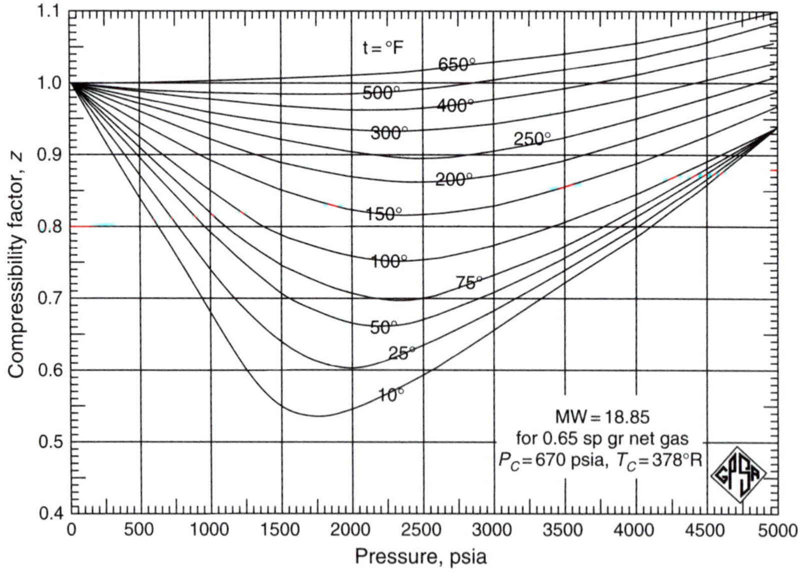

FIGURE 1.3. Compressibility factor for specific gravity = 0.65 gases (courtesy of GPSA Engineering Data Book).

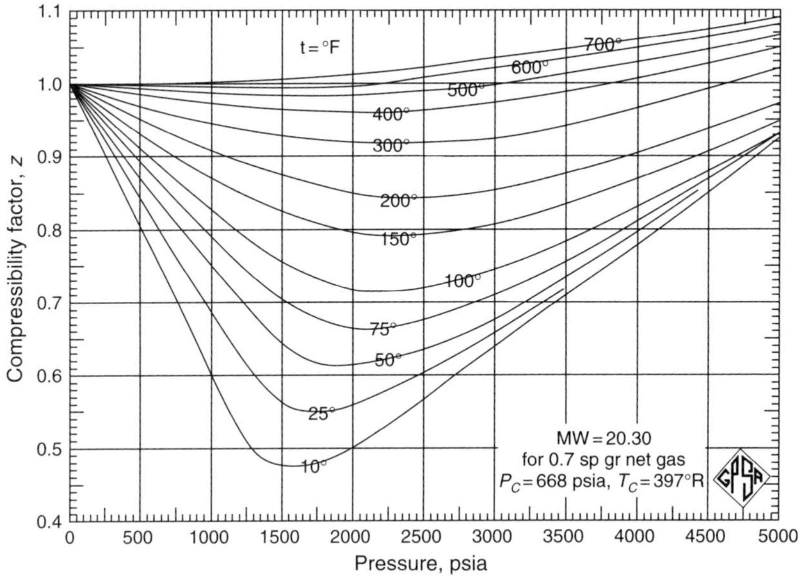

FIGURE 1.4. Compressibility factor for specific gravity = 0.7 gases (courtesy of GPSA Engineering Data Book).

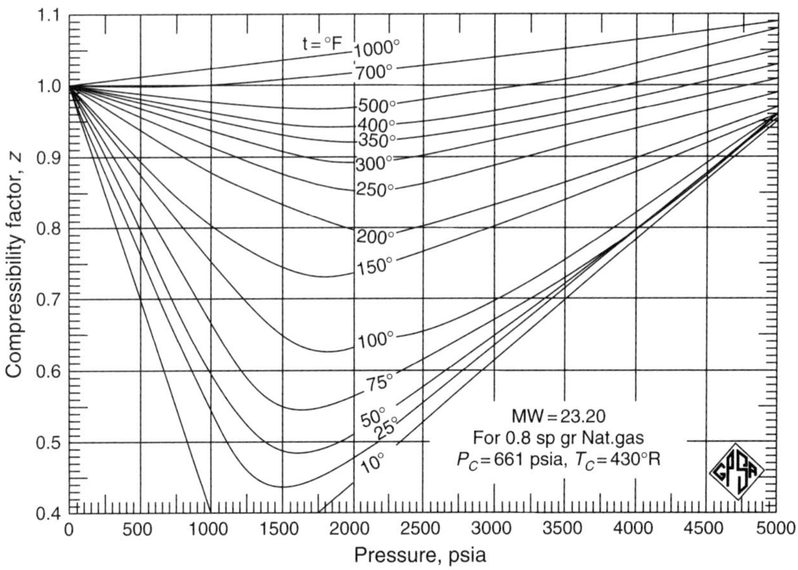

FIGURE 1.5. Compressibility factor for specific gravity = 0.8 gases (courtesy of GPSA Engineering Data Book).

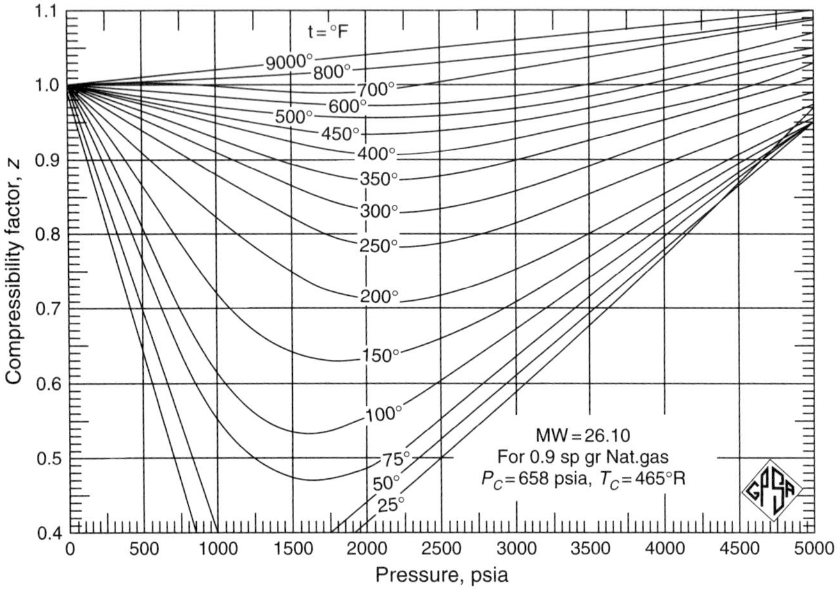

FIGURE 1.6. Compressibility factor for specific gravity = 0.9 gases (courtesy of GPSA Engineering Data Book).

where

SG = specific gravity of liquid
ρ_l = density of liquid
ρ_w = density of water at 60°F

The density of crude oil is sometimes shown in °API. This term is defined by the equation

$$SG = \frac{141.5}{131.5 + {}^\circ API} \tag{1.11}$$

or

$$^\circ API = \frac{141.5}{SG} - 131.5 \tag{1.12}$$

In most calculations, the specific gravity of liquids is normally referenced to actual temperature and pressure conditions. Figure 1.7 can be used to approximate how the specific gravity of a liquid decreases with increasing temperature, assuming no phase changes. In most practical pressure drop calculations associated with production facilities, the difference in specific gravity caused by pressure changes will not be severe enough to be considered if there are no phase changes.

For hydrocarbons, which undergo significant phase changes, Figure 1.8 can be used as an approximation of the specific gravity at a given pressure and temperature, once the API gravity of the liquid is known.

It should be pointed out that both Figures 1.7 and 1.8 are approximations only for the liquid component. Where precise calculation is required for a hydrocarbon, it is necessary to consider the gas that is liberated with decreasing pressure and increasing temperature. Thus, if a hydrocarbon is heated at constant pressure, its specific gravity will increase as the lighter hydrocarbons are liberated. The change in the molecular makeup of the fluid is calculated by "flash calculation," which is described in more detail later in this chapter.

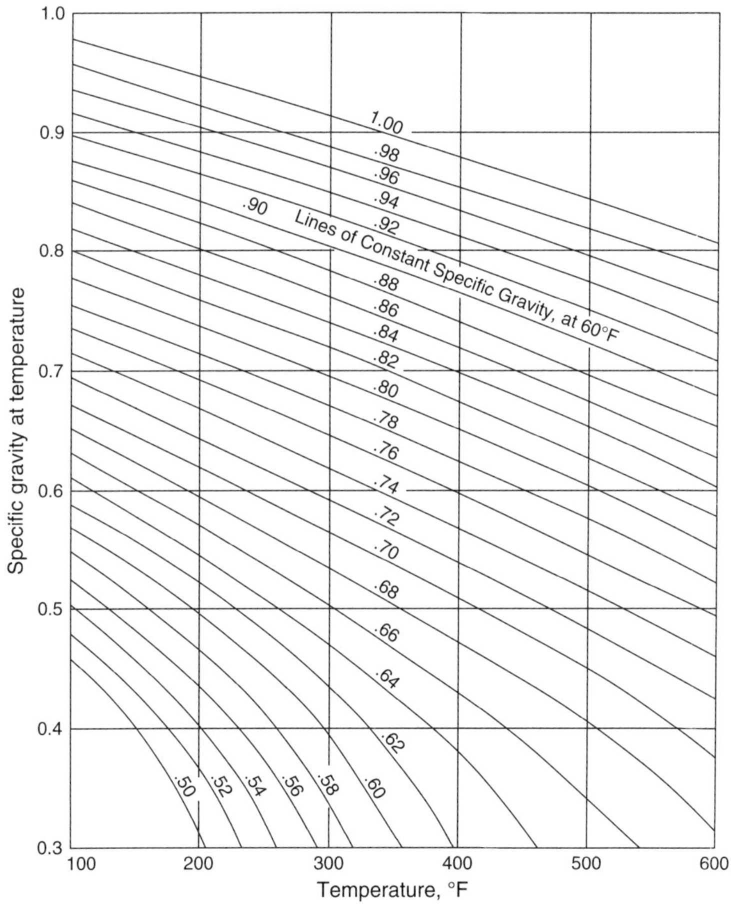

FIGURE 1.7. Approximate specific gravity of petroleum fractions (courtesy of GPSA Engineering Data Book).

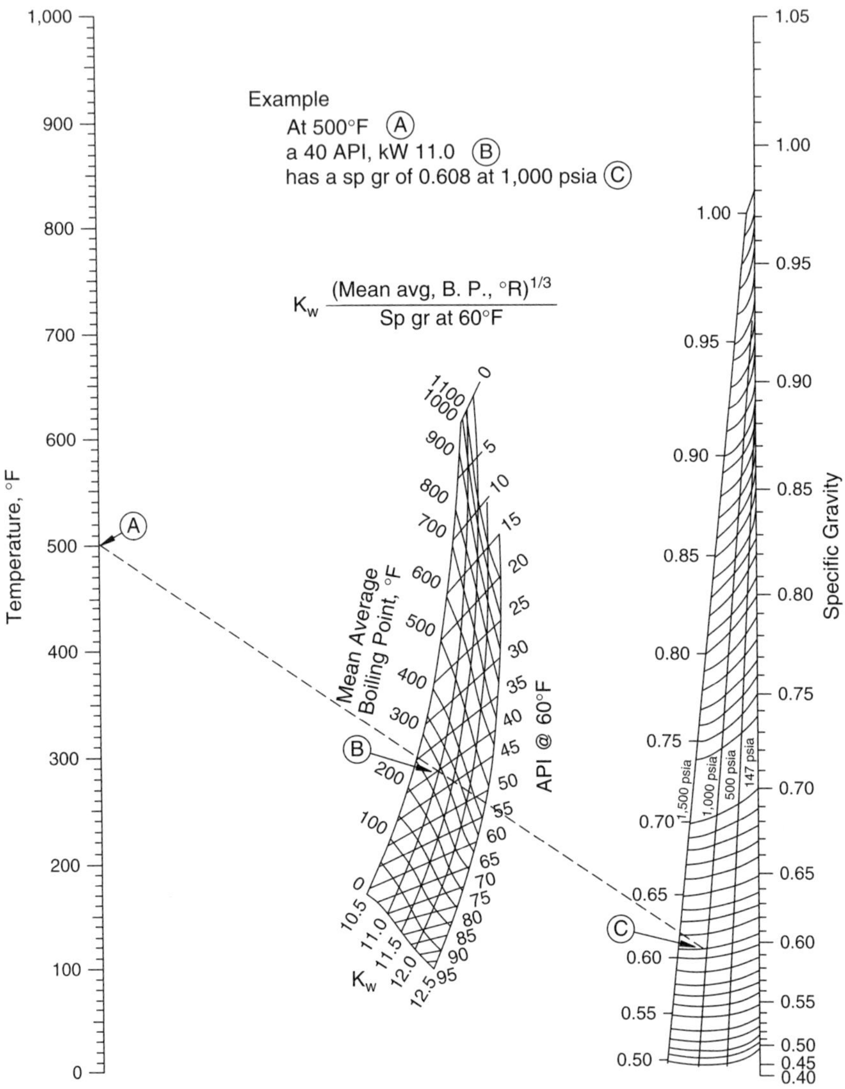

FIGURE 1.8. Specific gravity of petroleum fractions (courtesy of Petroleum Refiner: Ritter, Lenory, and Schweppe 1958).

1.3.6 Liquid Volume

By definition, 1 API barrel = 42 U.S. gallons at 60 °F

1 API bbl = 42 U.S. gallons = 35 U.K. (Imperial) gallons = 5.61 ft^3 = 0.159 m^3 = 159l

1.3.7 Viscosity

This property of a fluid indicates its resistance to flow. It is an important property used in flow equations and sizing of process equipment. It is a dynamic property in that it can be measured only when the fluid is in motion. Viscosity is a number that represents the drag forces caused by the attractive forces in adjacent fluid layers. It might be considered as the internal friction between molecules, separate from that between the fluid and the pipe wall.

$$1 \text{ centiPoise (cP)} = 0.01 \text{ dyn s/cm}^2 = 0.000672 \text{ lb m/ft s}$$

There are two expressions of viscosity: absolute (or dynamic) viscosity, μ, and kinematic viscosity.

These expressions are related by the following equation:

$$Y = \frac{\mu}{\rho} \tag{1.13}$$

where

μ = absolute viscosity, cP
Y = kinematic viscosity, centistokes (cSt)
ρ = density, g/cm^3
and
$1 \text{ cSt} = 0.01 \text{ cm}^2/\text{sec} = 1.0 \times 10^{-6} \text{ m}^2/\text{sec}$

Fluid viscosity changes with temperature. Liquid viscosity decreases with increasing temperature, whereas gas viscosity decreases initially with increasing temperature and then increases with further increasing temperature.

Figure 1.9 can be used to estimate the viscosity of a hydrocarbon gas at various conditions of temperature and pressure if the specific gravity of the gas at standard conditions is known. It is useful when the gas composition is not known. It does not make corrections for H_2S, CO_2, and N_2. It is useful for determining viscosities at high pressure. Unfortunately, it is an approximate correlation and thus yields less accurate results than other correlations, but for most engineering calculations Figure 1.9 yields results within acceptable limits. When compared to liquid viscosity, gas viscosity is very low, which indicates the relatively large distances between molecules.

The best way to determine the viscosity of a crude oil at any temperature is by measurement. If the viscosity is not known, Figure 1.10 can be used as a rough approximation. If the viscosity is known at only one temperature, Figure 1.10 can be used to determine the viscosity at another temperature by striking a line parallel to the lines shown. Care must be taken to assure that the crude does not have its pour point within the temperature range of interest. If it does,

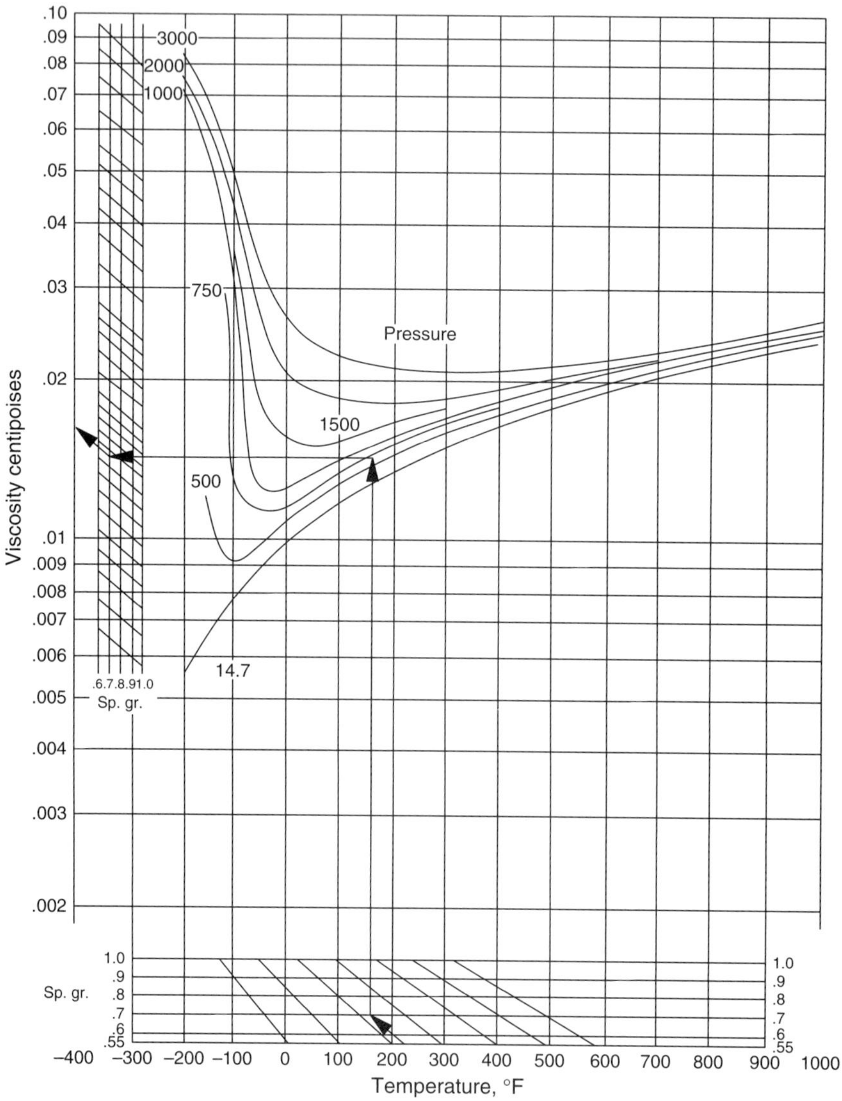

FIGURE 1.9. Hydrocarbon gas viscosity (courtesy of GPSA Engineering Data Book).

its temperature–viscosity relationship may be as shown for crude "B" in Figure 1.11.

Solid phase high-molecular-weight hydrocarbons, otherwise known as paraffins, can dramatically affect the viscosity of the crude sample. The cloud point is the temperature at which paraffins first

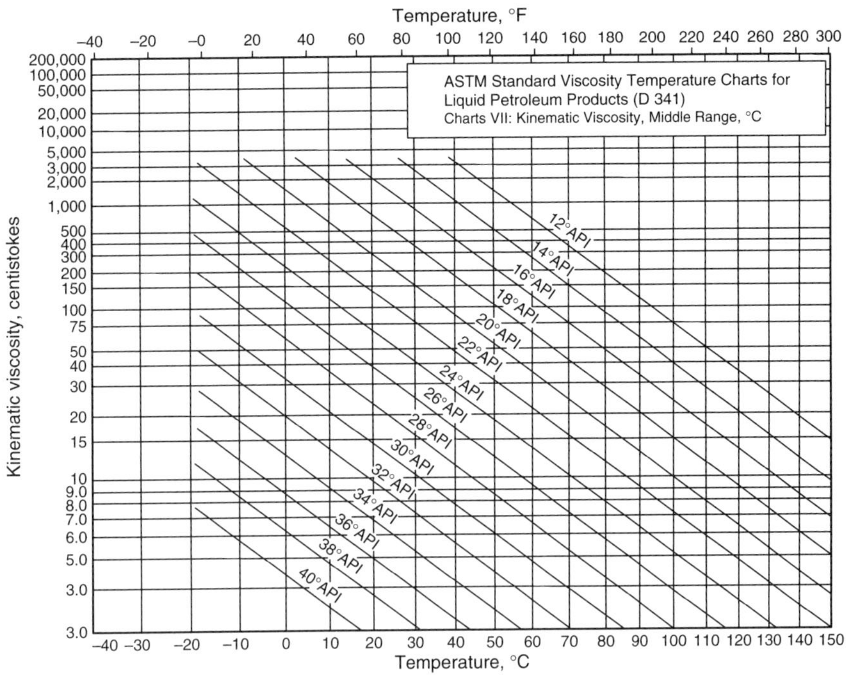

FIGURE 1.10. Oil viscosity vs. gravity and temperature (courtesy of Paragon engineering Services, Inc.).

become visible in a crude sample. The effect of the cloud point on the temperature–viscosity curve is shown for crude "B" in Figure 1.11. This change in the temperature–viscosity relationship can lead to significant errors in estimation. Therefore, care should be taken when one estimates viscosities near the cloud point.

The pour point is the temperature at which the crude oil becomes a solid and ceases to flow, as measured by a specific ASTM procedure (D97). Estimations of viscosity near the pour point are highly unreliable and should be considered accordingly.

The viscosity of produced water depends on the amount of dissolved solids in the water as well as the temperature, but for most practical situations it varies from 1.5 to 2 cP at 50 °F, 0.7 to 1 cP at 100 °F, and 0.4 to 0.6 cP at 150 °F.

When an emulsion of oil and water is formed, the viscosity of the mixture may be substantially higher than either the viscosity of the oil or that of the water taken by themselves. Figure 1.12 shows some experimental data for a mixture of produced oil and water taken from a south Louisiana field. Produced oil and water were mixed vigorously

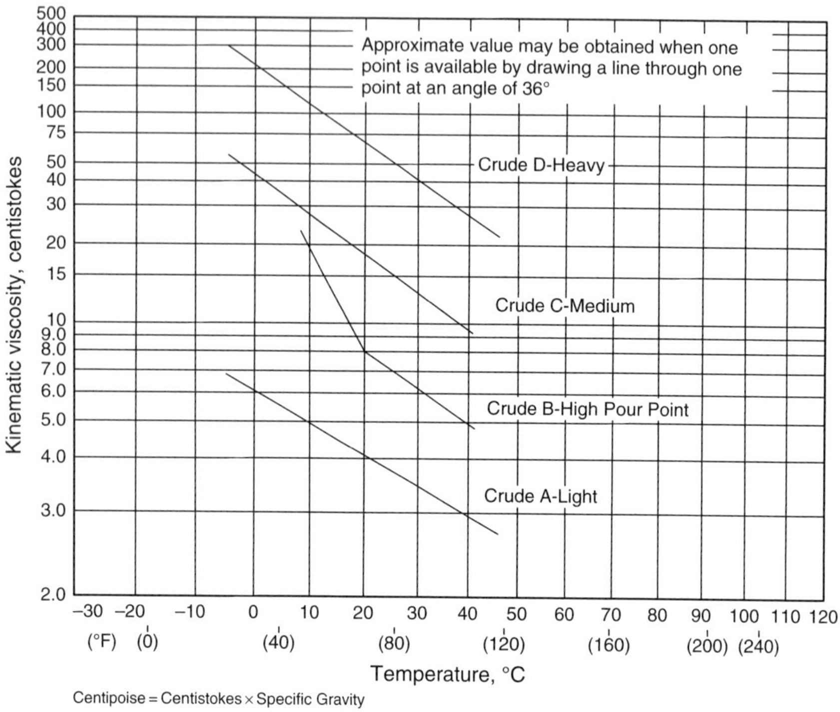

FIGURE 1.11. Typical viscosity–temperature curves for crude oils (courtesy of ASTM D-341).

by hand, and viscosity was measured for various percentages of water. For 70% water cut, the emulsion began to break before viscosity readings could be made, and for water cuts greater than this, the oil and water began to separate as soon as the mixing stopped. Thus, at approximately 70% water cut, it appears as if oil ceases to be the continuous phase and water becomes continuous.

The laboratory data plotted in Figure 1.12 agree closely with the modified Vand's equation assuming a 70% breakover point. This equation allows one to determine the effective viscosity of an oil–water mixture and is written in the form

$$\mu_{\text{eff}} = (1 + 2.5\emptyset + 10\emptyset^{2})\mu_{\text{c}} \tag{1.14}$$

where

μ_{eff} = effective viscosity
μ_{c} = viscosity of the continuous phase
$\emptyset$ = volume fraction of the discontinuous phase

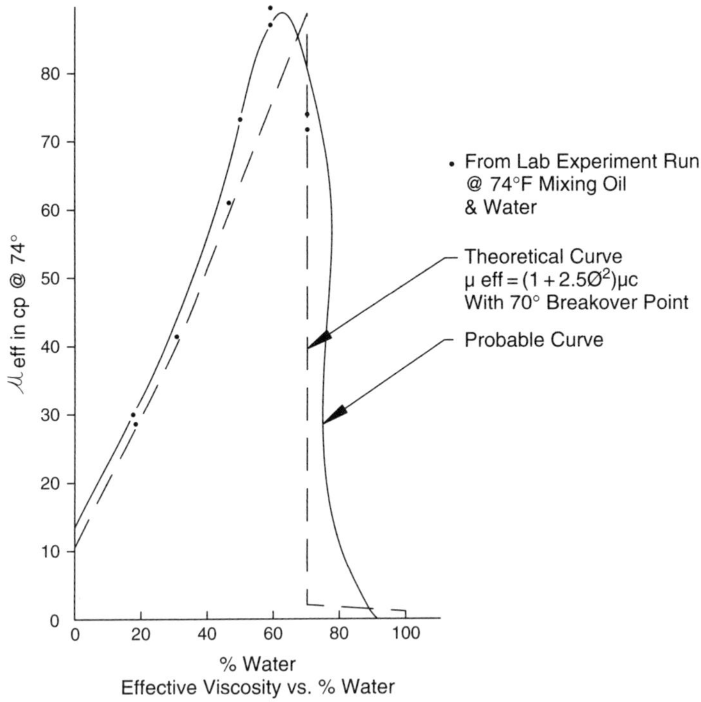

FIGURE 1.12. Effective viscosity of an example oil/water mixture.

1.4 Flash Calculations

1.4.1 Determine Gas and Liquid Compositions

The amount of hydrocarbon fluid that exists in the gaseous phase or the liquid phase at any points at the process is determined by a flash calculation. As explained in Chapter 2 (this volume), for a given pressure and temperature, each component in the gas phase will depend not only on pressure and temperature, but also on the partial pressure of the component. Therefore, the amount of gas depends upon the total composition of the fluids as the mole fraction of any one component in the gas phase is the function of the mole fraction of every other component in this phase.

This is best understood by assigning an equilibrium "*K*" value to each component. The *K* value is a strong function of temperature and pressure and of the composition of the vapor and liquid phase. It is defined as

$$K_N = \frac{V_N/V}{L_N/L} \tag{1.15}$$

where

K_N = constant for component N at a given temperature and pressure

V_N = moles of component N in the vapor phase

V = total moles in the vapor phase

L_N = moles of component N in the liquid phase

L = total moles in the liquid phase

The Gas Processors Suppliers Association (GPSA) present graphs of K values for the important components in a hydrocarbon mixture such as that shown in Figure 1.13. The K values are for specific "convergence" pressure. A procedure in the GPSA's *Engineering Data Book* for calculating convergence pressure based on simulating a binary fluid system with the lightest hydrocarbon component, which makes up at least 0.1 mol% in the liquids and a pseudo-heavy component having the same average weight and critical temperature as the remaining heavier hydrocarbons. The convergence pressure is then read from a graph of convergence pressure versus operating temperature for common pseudo-binaries.

In most oil-field applications the convergence pressure will be between 2000 and 3000 psia, except at very low pressures, where a psia between 500 and 1500 is possible. If the operating pressure is much less than the convergence pressure, the equilibrium constant is not greatly affected by the choice of convergence pressure. Therefore, using a convergence pressure of 2000 psia is a good first approximation for most flash calculations. Where greater precision is required, the convergence pressure should be calculated.

If K_N for each component and the ratio of total moles of vapor to total moles of liquid (V/L) are known, then the moles of the component N in vapor phase (V_N) and the moles in the liquid phase (L_N) can be calculated from

$$V_N = \frac{K_N F_N}{\frac{1}{V/L} + K_N} \tag{1.16}$$

$$L_N = \frac{F_N}{K_N(V/L) + 1} \tag{1.17}$$

where F_N = total moles of component N in the fluid.

To solve either Equation (1.16) or (1.17), it is necessary to first know the quantity (V/L), but since both V and L are determined by summing V_N and L_N, it is necessary to use an iterative solution. This is done by estimating (V/L), calculating V_N and L_N for each component, summing up to obtain the total moles of gas (V) and liquid (L), and then comparing the calculated (V/L) to assumed value.

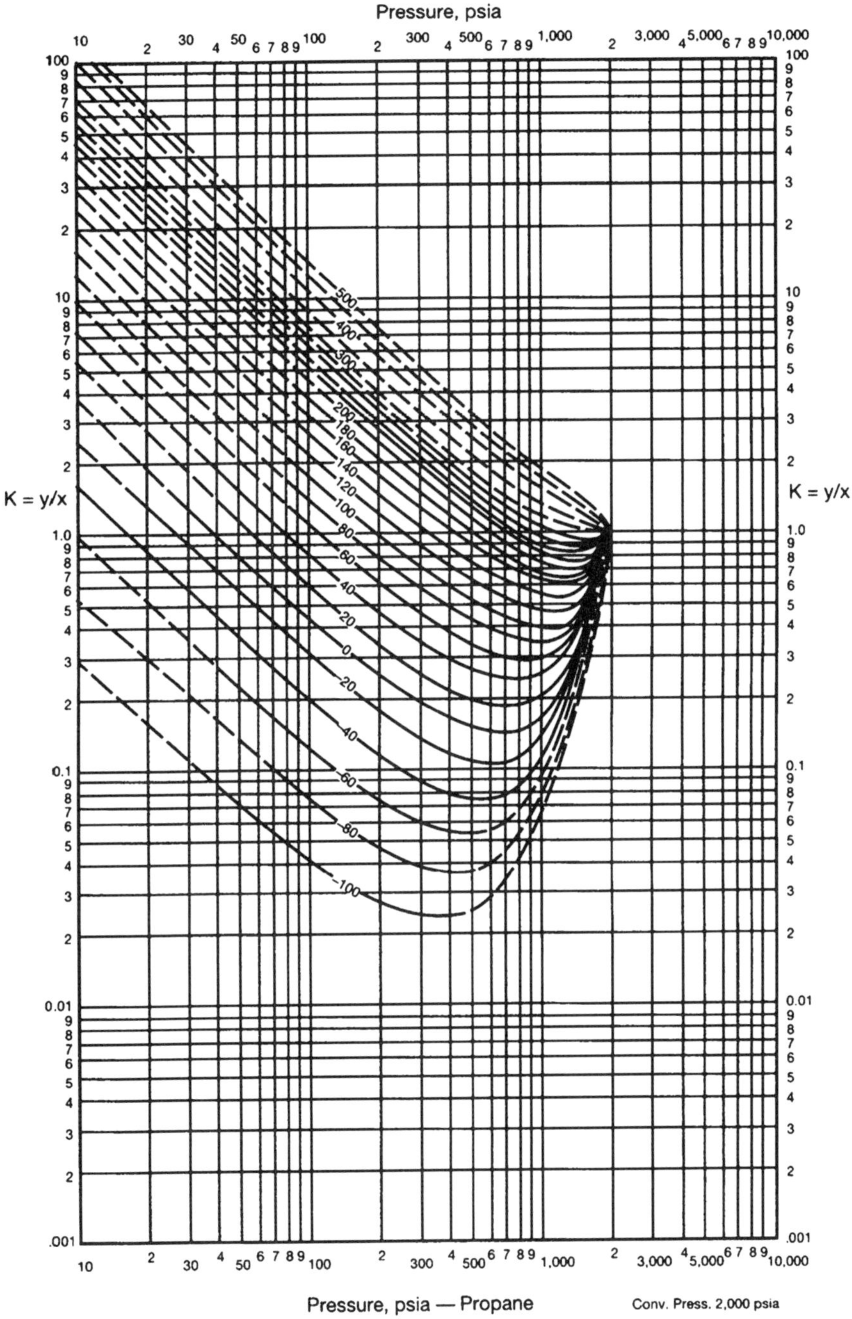

FIGURE 1.13. "K" values for propane (courtesy of GPSA Engineering data book).

1.4.2 Characterizing the Flow Stream

Once a flash calculation is made and the molecular composition of the liquid and gas components have been determined, it is possible to determine the properties and flow rates of both the gas and the liquid streams.

The molecular weight of a stream is calculated from the weighted average gas molecular weight given by

$$\mathrm{MW} = \sum [V_N \times (\mathrm{MW})_N] \tag{1.18}$$

The gas's specific gravity can be determined from the molecular weight from Equation (1.7). If the flow of the inlet stream is known in moles per day, then the number of moles per day of gas flow can be determined from

$$V = \frac{F}{1 + \dfrac{1}{V/L}} \tag{1.19}$$

where

V = gas flow rate, mol/day
F = total stream flow rate, mol/day
L = liquid flow rate, mol/day

Once the mole flow rate of gas is known, then the flow rate in standard cubic feet can be determined by recalling that one mole of gas occupies 380 ft^3 at standard conditions. Therefore,

$$Q_g = \frac{380V}{1{,}000{,}000} \tag{1.20}$$

where Q_g = gas flow rate, MMscfd.

The molecular weight of the liquid stream is calculated from the weighted average liquid component molecular weight given by

$$\mathrm{MW} = \frac{\sum [L_N \times (\mathrm{MW})_N]}{L} \tag{1.21}$$

Remembering that the weight of each component is the number of moles of that component times its molecular weight, the specific gravity of the liquid is given by

$$\mathrm{SG} = \frac{\sum [L_N \times (\mathrm{MW})_N]}{\sum \dfrac{[L_N \times (\mathrm{MW})]_N}{(\mathrm{SG})_N}} \tag{1.22}$$

The liquid flow rate in barrels per day can be derived from

$$Q_l = \frac{L \times (\mathrm{MW})}{350(\mathrm{SG})}, \tag{1.23}$$

where
Q_l = liquid flow rate, bpd
SG = specific gravity of liquid (water = 1).

Many times the designer is given the mole fraction of each component in the feed stream but is not given the mole flow rate for the stream. It may be necessary to estimate the total number of moles in the feed stream (F) from an expected stock-tank oil flow rate. As a first approximation, it can be assumed that all the oil in the stock tank can be characterized by the C_7^+ component of the stream. Thus, the feed rate in moles per day can be approximated as

$$L \cong \frac{350(\mathrm{SG})_7 Q_l}{(\mathrm{MW})_7}, \tag{1.24}$$

where
L = liquid flow rate, mol per day,
$(\mathrm{SG})_7$ = specific gravity of C_7^+,
$(\mathrm{MW})_7$ = molecular weight of C_7^+,
Q_l = flow rate of liquid, bpd.

The mole flow rate of the feed stream is then calculated as

$$F = \frac{L}{(\text{mole fraction})_7} \tag{1.25}$$

where
F = flow rate feed stream, mol/day
$(\text{mole fraction})_7$ = mole fraction of the C_7^+ component in the feed stream.

The flash calculation could then proceed. The calculated flow rates for each stream in the process could then be used in a ratio to reflect the error between assumed stock-tank flow rate and desired stock-tank flow rate.

Refer to Surface Production Operations, Volume 1, pages 135–136, for a complete example using this hand calculation method.

1.5 Use of Computer Programs for Flash Calculations

The iterative manual flash calculation detailed in the previous sections shows one of many methods for calculating equilibrium conditions. Flash calculations are inherently rigorous and best performed by sophisticated simulation software, such as HYSIM or other similar programs.

1.6 Approximate Flash Calculations

Sometimes it is necessary to get a quick estimate of the volume of gas that is expected to be flashed from an oil stream at various pressures.

Figure 1.14 was developed by flashing several crude oils of different gravities at different pressure ranges. The curves are approximate. The actual shape would depend on the initial separation pressure, the number and pressure of intermediate flashes, and the temperature.

Use of the curve is best explained by an example. Suppose a 30 × API crude with a GOR of 500 is flashed at 1000 psia, 500 psia, and 50 psia before going to a stock-tank. Roughly 50% of the gas that will eventually be flashed from the crude, or 250 ft^3/B, will be liberated as gas in the 1000-psia separator. Another 25% (75–50%), or 125 ft^3/B, will be separated at 500 psia, and 23% (98%–75%), or 115 ft^3/B, will be separated at 50 psia. The remaining 10 ft^3/B (100–98%) will be vented from the stock tank.

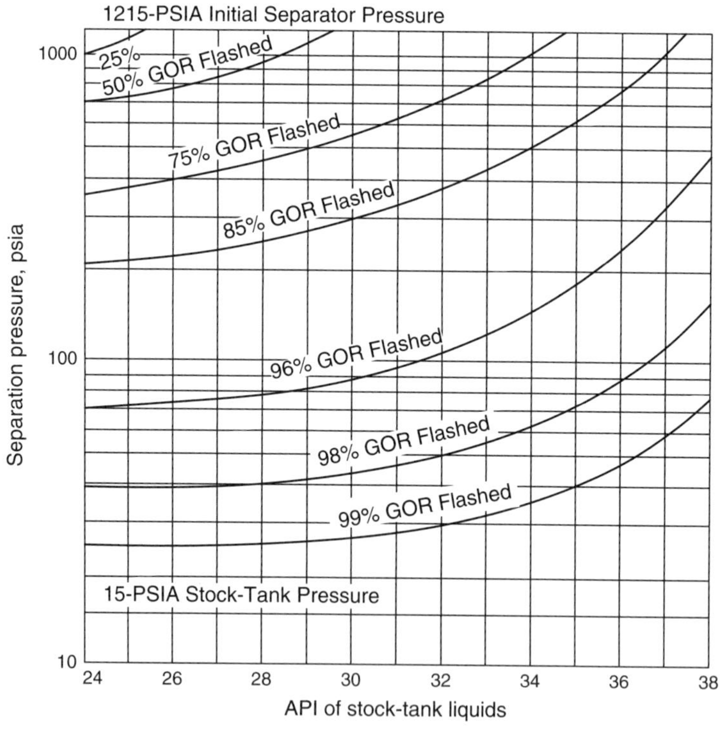

FIGURE 1.14. Preliminary estimation of % GOR flashed for given API of stock tank liquids and separation pressures-Gulf Coast Crudes.

It must be stressed that Figure 1.14 is only to be used where a quick approximation, which could be subject to error, is acceptable. It cannot be used for estimating gas flashed from condensate produced in gas wells.

1.7 Other Properties

Once the equilibrium conditions (and, therefore, the gas and the liquid compositions) are known, several very useful physical properties are obtainable, such as the dew point, the bubble point, the heating value (net and gross), and k, the ratio of gas-specific heats. These properties are described next:

Dew point: the point at which liquid first appears within a gas sample.

A more precise definition of the dew point makes a distinction between the hydrocarbon dew point, which represents the condensation of a hydrocarbon liquid, and the water dew point, which represents the condensation of liquid water. Often, sales gas contracts specify control of the water dew point for hydrate and corrosion control and not the hydrocarbon dew point. In such cases, hydrocarbons will often condense in the pipeline as the gas cools (assuming that separation has occurred at a higher temperature than ambient), and provisions to separate this "condensate" must be provided.

Bubble point: the point at which gas first appears within a liquid sample.

Net heating value: heat released by combustion of gas sample with water vapor as a combustion product; also known as the lower heating value (LHV).

Gross heating value: heat released by combusting of gas sample with liquid water as a combustion product; also known as the higher heating value (HHV).

k: ratio of heat capacity at constant pressure (C_P) to heat capacity at constant volume (C_V). Often used in compressor calculation of horsepower requirement and volumetric efficiencies. This ratio is relatively constant for natural gas molecular weight and ranges between 1.2 and 1.3 (see Figure 1.15).

Reid vapor pressure: the bubble point can also be referred to as the "true vapor pressure." A critical distinction lies here between the true vapor pressure and the Reid vapor pressure (RVP). The Reid vapor pressure is measured according to a specific ASTM standard (D323) and lies below the rue vapor pressure.

The approximate relationship between the two pressures is shown in Figure 1.16. (Note that an RVP below atmospheric pressure

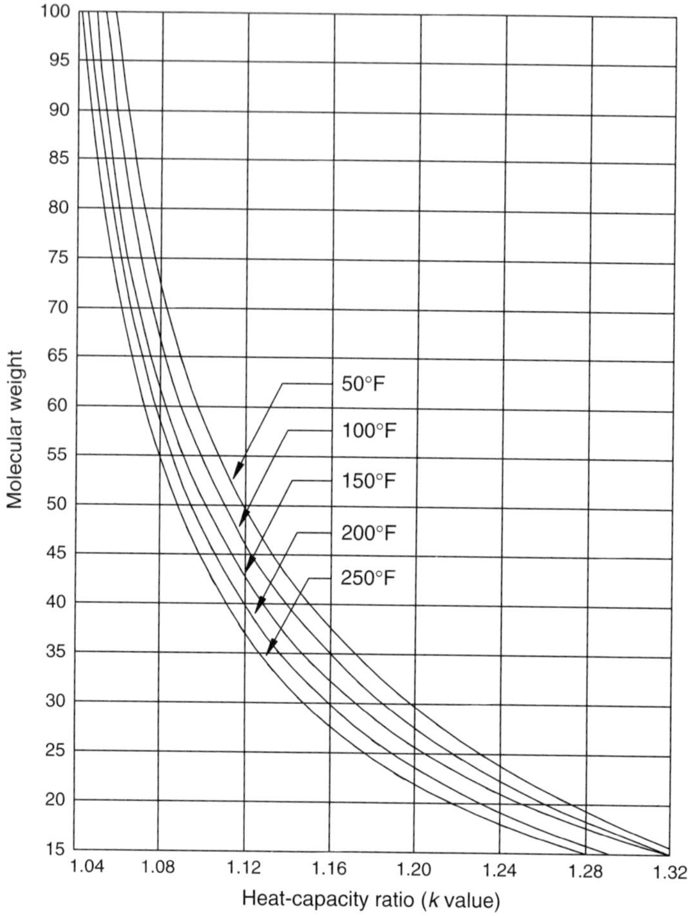

FIGURE 1.15. Approximate heat-capacity ratios of hydrocarbon gases (courtesy of GPSA Engineering data book).

does not indicate that vapors will be absent from a sample at atmospheric pressure.)

1.8 Phase Equilibrium

A basic representation of the equilibrium information for a specific fluid composition can be found in a P–H (pressure–enthalpy) diagram, which is highly dependent on the sample composition. This diagram can be used to investigate thermodynamic fluid properties as well as their thermodynamic phenomena such as retrograde condensation

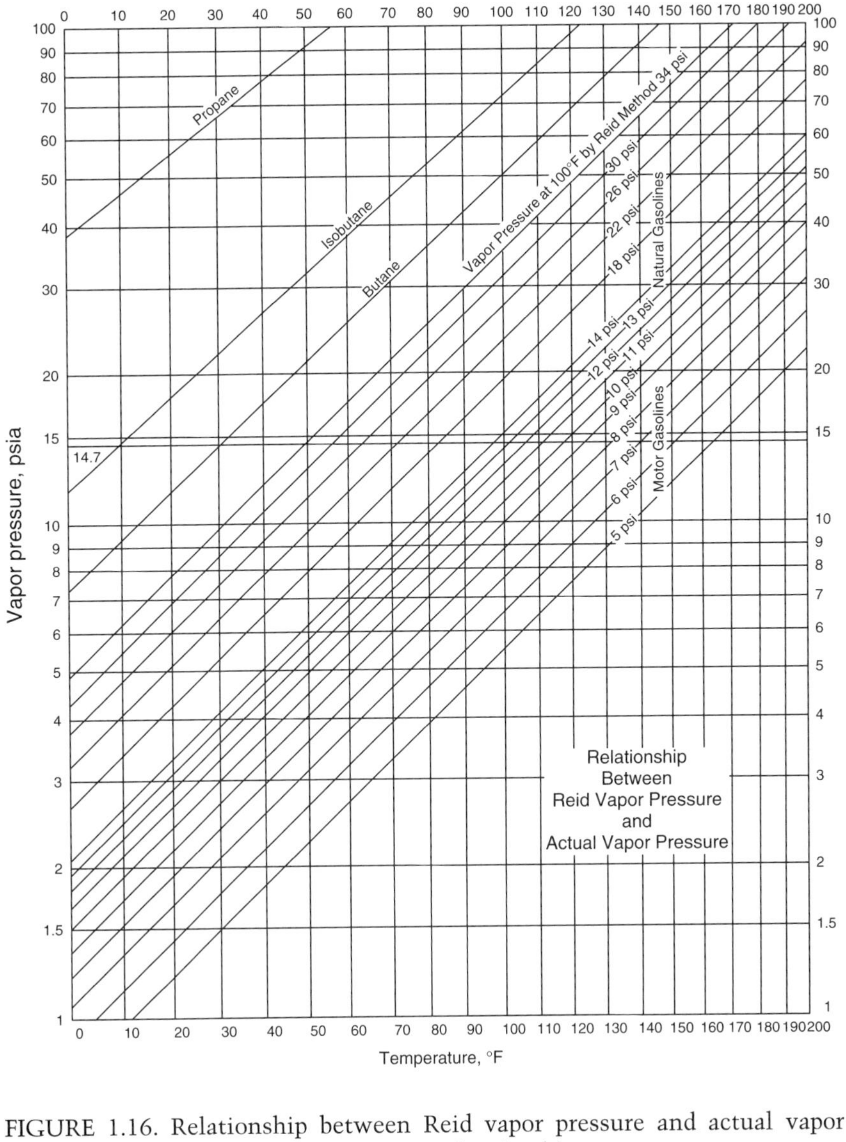

FIGURE 1.16. Relationship between Reid vapor pressure and actual vapor pressure (courtesy of GPSA Engineering data book).

and the Joule–Thomson effect. Please note, however, that a P–H diagram is unlikely to be available for anything but a single component of the mixture, unless the diagram is created by simulation software packages such as those mentioned above. A P–H diagram for propane is shown in Figure 1.17; a P–H diagram for a 0.6 specific gravity natural gas is shown in Figure 1.18.

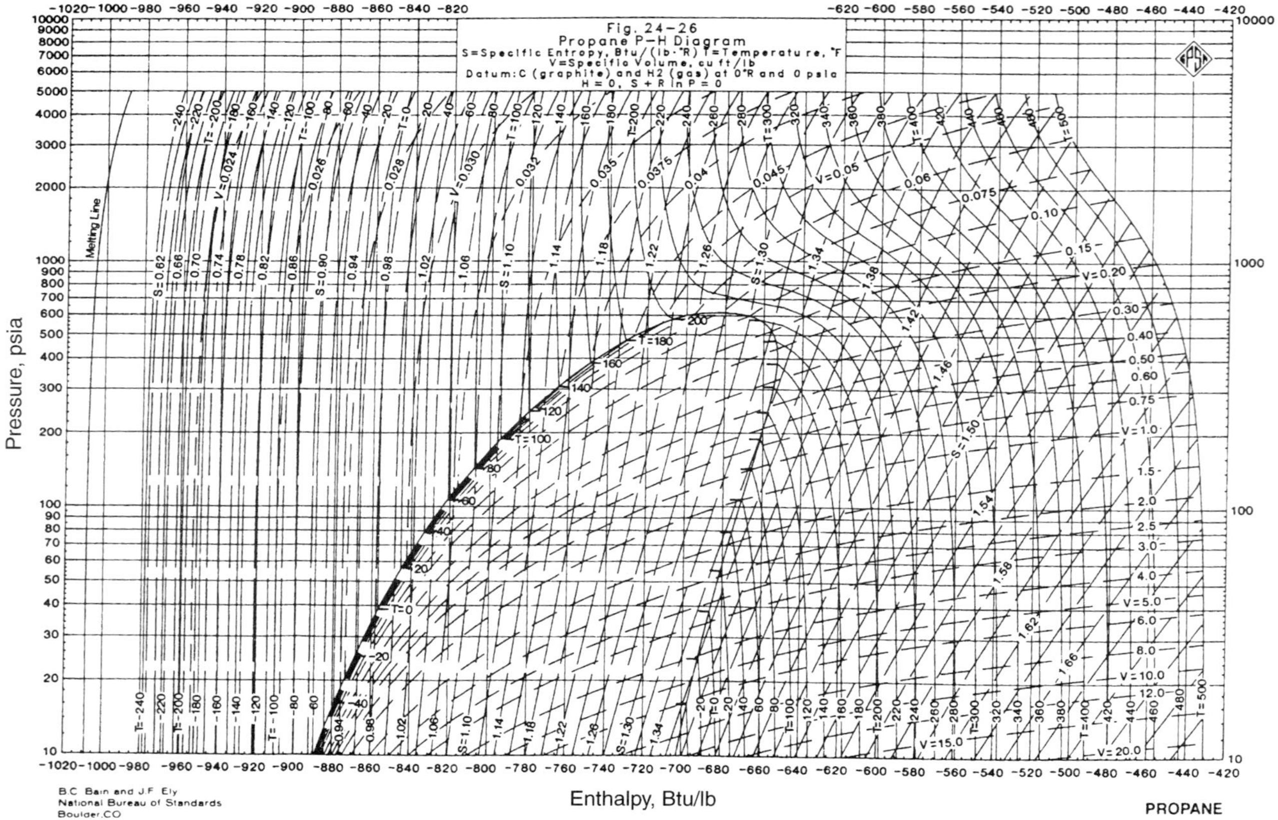

FIGURE 1.17. A P–H diagram for propane (courtesy of GPSA Engineering data book).

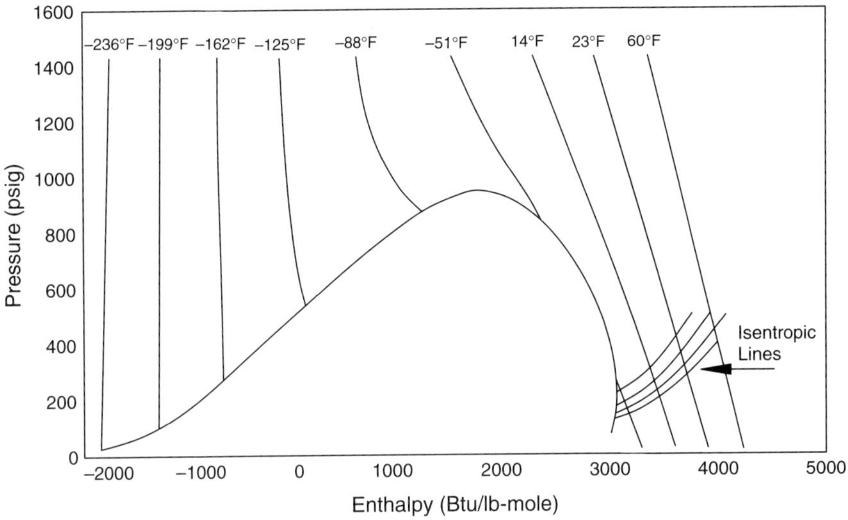

FIGURE 1.18. A P–H diagram for 0.6 specific gravity natural gas.

CHAPTER 2

Process Selection

2.1 Introduction to Field Facilities

This chapter

- provides an overview of the more detailed sections that follow and
- illustrates how the various components are combined into a production system.

Specifically, this chapter discusses the

- gathering, separation, and treating of crude oil for sale and refining;
- gathering, separation, compression, and treating of associated gas and condensate; and
- the treating and disposal of contaminants, such as water and solids.

Material is in no way meant to be all-inclusive. Many things must be considered in selecting components for a production system, and there is no substitute for experience and good engineering judgment. Process flowsheet/diagram (PFD), shown in Figure 2.1, is used to describe the production system. Figure 2.2 defines many of the commonly used symbols in PFDs.

We begin with controlling the process followed by a description of the reservoir fluid characteristics. Remaining sections contain a brief overview of

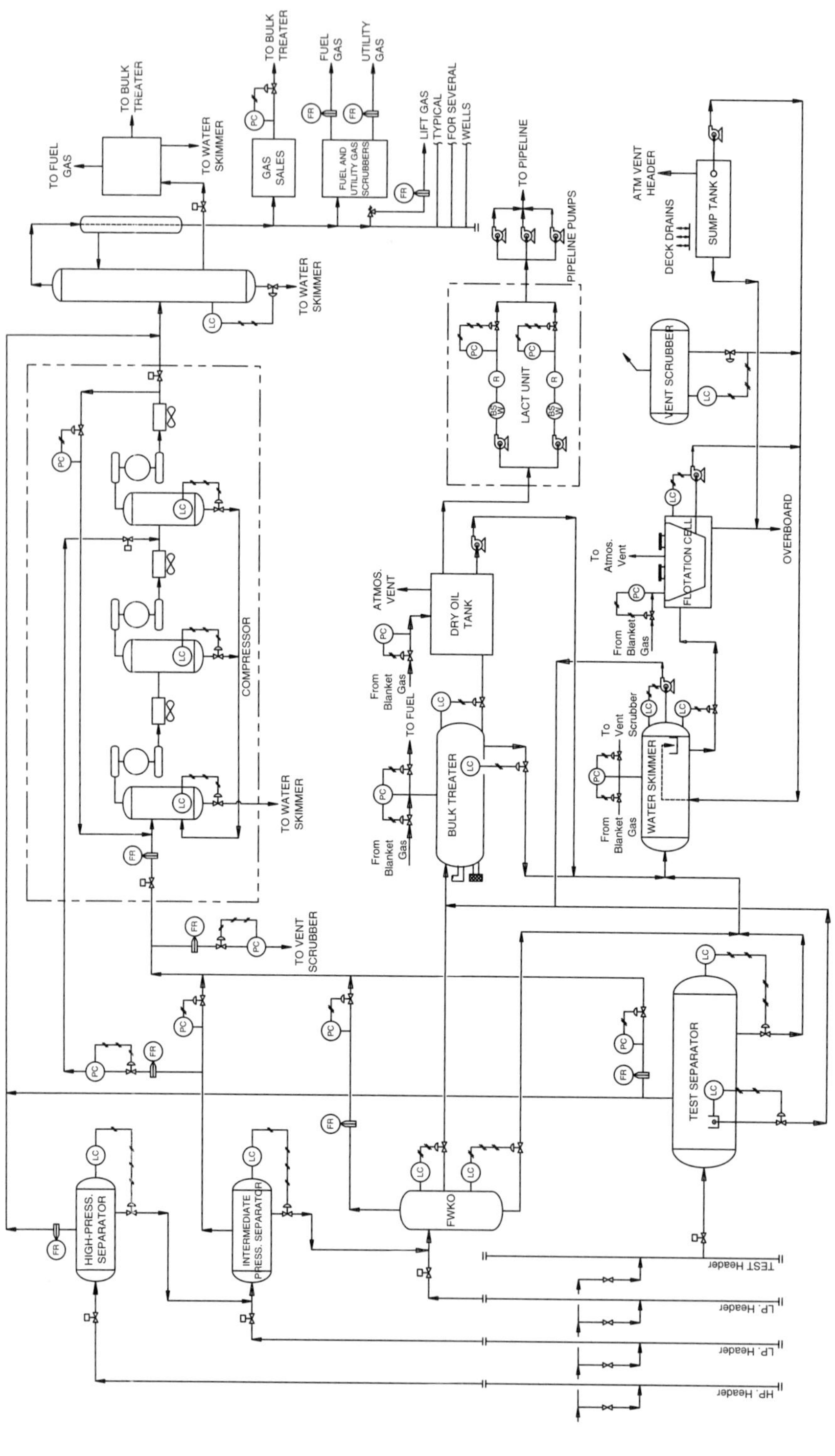

FIGURE 2.1. Typical flowsheet.

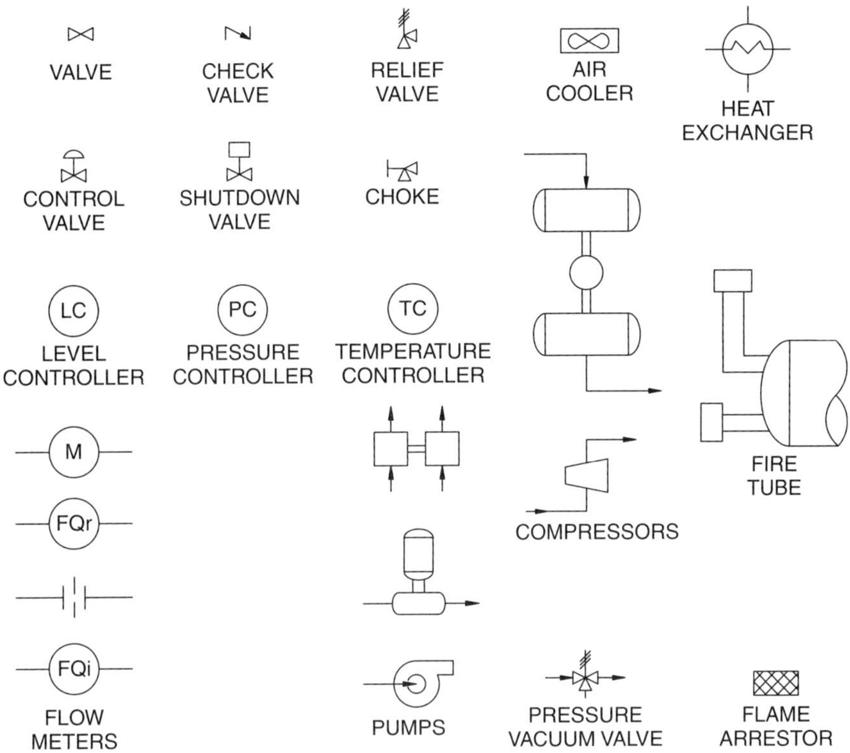

FIGURE 2.2. Common flowsheet symbols.

- basic system configuration, including the equipment, facilities, and processes typically encountered in oil and gas production operations, and
- well testing, gs lift, and offshore platform considerations.

Before discussing the process itself, it is necessary to understand how the process is controlled.

2.2 Controlling the Process

2.2.1 Operation of a Control Valve

Control valves are used throughout the process to control

- pressure,
- level,
- temperature, and
- flow.

Discussion about the various types of control valves and sizing procedures are beyond the scope of this chapter. These topics are discussed in detail in another volume of the series.

All control valves have a variable opening or orifice. For a given pressure drop across the valve, the larger the orifice, the greater the flow through the valve. Chokes and other flow control devices have either a fixed or a variable orifice. For a fixed pressure drop across the device (i.e., with both the upstream and downstream pressures fixed by the process system), the larger the orifice, the greater the flow through the valve. Chokes are used to regulate the flow rate.

Figure 2.3 shows the major components of a typical sliding-stem control valve. The orifice is made larger or smaller by moving the valve stem upward or downward. Moving the valve stem upward creates a larger annulus for flow between the seat and the plug. Moving the stem downward creates a smaller annulus and less flow.

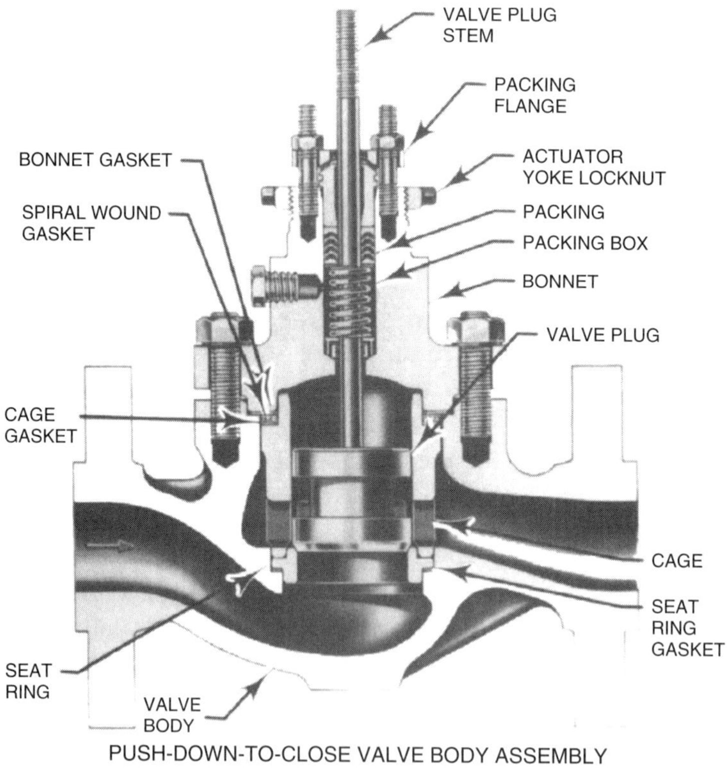

FIGURE 2.3. Major components of a typical sliding-stem control valve (courtesy of Fisher Controls International, Inc.).

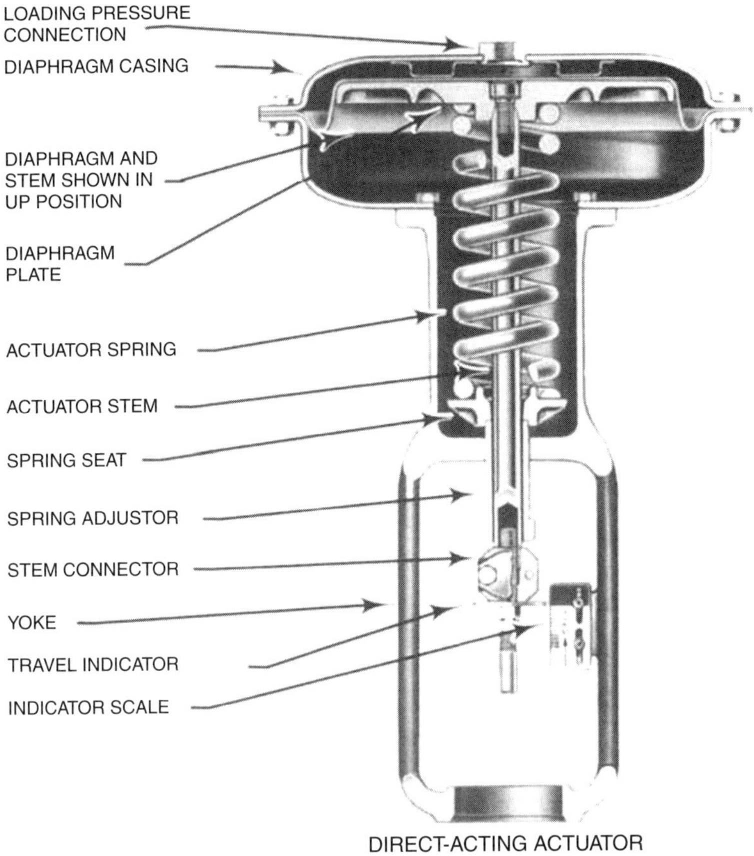

FIGURE 2.4. Typical pneumatic direct-acting actuator (courtesy of Fisher Controls International, Inc.).

The most common way to effect this motion is with a pneumatic actuator. Figure 2.4 shows a typical pneumatic direct-acting actuator. Instrument air or gas applied to the actuator diaphragm overcomes a spring resistance and moves the stem either upward or downward. The action of the actuator must be matched with the construction of the valve body to ensure that the required failure mode is met. If it is desirable for the valve to fail close, as in many liquid dump valves, then the actuator and valve body must be matched so that on failure of the instrument air or gas, the spring causes the stem to move in the direction that blocks flow (i.e., fully shut). If it is desirable for the valve to fail open, as in many pressure control situations, then the spring must cause the stem to move in the fully open direction.

2.2.2 Pressure Control

Well fluids are made up of many components ranging from methane—the lightest—to very heavy and complex compounds. Whenever there is a pressure drop in fluid pressure, gas is liberated and thus pressure control is important. Pressure is normally controlled with a pressure controller and a backpressure control valve. Pressure controller senses the pressure in the vapor space of the vessel or tank. Backpressure control valve maintains the desired pressure in the vessel by regulating the amount of gas leaving the vapor space.

If too much gas is liberated, the number of gas molecules in the vapor space decreases, and thus the pressure in the vessel decreases. If too little gas is liberated, the number of gas molecules in the vapor space increases, and thus the pressure in the vessel increases.

In most instances, there is sufficient gas separated, or "flashed," from the liquid to allow the pressure controller to compensate for changes in liquid level, temperature, and so on, which would cause a change in the number of molecules of gas required to fill the vapor space at a given pressure.

Pressure is sometimes controlled by adding "Makeup" or "Blanket" gas to the vessel—used where there is a small pressure drop from the upstream vessel or where the crude GOR (gas/oil ratio) is low. Gas from a higher-pressure source is routed to the vessel by a pressure controller that senses the vessel pressure automatically, allowing either more or less gas to enter the vessel as required.

2.2.3 Level Control

Level controller and dump valve is used to control the gas/liquid interface and/or the oil/water interface. Most common forms of level controllers include floats, displacers, and electronic sensing devices. The controller and dump valves are constantly adjusting its opening to ensure that the rate of liquid flowing into the vessel is matched by the rate out of the vessel. If the level begins to rise, the controller signals the liquid dump valve to open and allow liquid to leave the vessel. If the level begins to fall, the controller signals the liquid dump valve to close and decrease the flow of liquid from the vessel.

2.2.4 Temperature Control

The way in which the process temperature is controlled varies. In a heater, a temperature controller measures the process temperature and signals a fuel valve to let either more or less fuel to the burner. In a heat exchanger, the temperature controller could signal a valve to allow more or less of the heating or cooling media to bypass the exchanger.

2.2.5 Flow Control

It is rare that flow must be controlled in an oil field process. Normally, the control of pressure, level, and temperature is sufficient to control flow. Occasionally, it is necessary to ensure that flow is split in some controlled manner between two process components in parallel or perhaps to maintain a certain critical flow through a component. This can become a complicated control problem and must be handled on an individual basis.

2.3 Reservoir Fluid Characteristics

Reservoir fluids

- are usually under high pressure,
- are in contact with water which is usually salty, and
- may be in a liquid or gaseous state.

Each reservoir is unique.
Individual characteristics will have an effect on

- how the wells will be produced and
- how they must be treated when they reach the surface.

Important reservoir fluid characteristics are

- size and shape,
- depth below the surface,
- type of rock that it consists of,
- pressure and temperature,
- type and quantity of fluid that it contains,
- whether the fluid contains components considered to be undesirable (i.e., H_2S or CO_2), and
- amount of dissolved solids in the water.

2.4 Basic System Configuration

2.4.1 Wellhead and Manifold

Production system begins at the wellhead, which includes a minimum of one choke, unless the well is on an artificial lift.

Choke

- Pressure upstream is determined by the well FTP (flowing tubing pressure).

- Pressure downstream is determined by the pressure control valve on the first separator in the system.
- Size of the opening determines the flow rate.

Multiple chokes

- Usually required on high-pressure wells.
- Incorporates a positive choke in series with an adjustable choke.
- Positive choke takes over and keeps the production rate within limits should the adjustable choke fail.

Automatic surface safety valve (SSV)

- Installed on high-risk installations.
- Required by the authorities having jurisdiction on all offshore facilities.

Isolation block valves

- Allows maintenance to be performed without having to shut-in the wellhead.

Manifold

- Required whenever two or more wells are commingled in a central facility.
- Allows flow from one well to be produced into any of the bulk or test systems.

2.4.2 Separation

General

When reservoir fluids reach the surface, they usually contain a mixture of gas, oil, and water (refer to Figure 2.5). Separation, which represents the first surface production step, separates these three fluids.

As shown in Figure 2.6, after initial separation, each stream is processed in a different manner. After the oil and gas have been treated to achieve a marketable quality, very accurate measurements are required for the purpose of custody transfer. Separation is often accomplished in two or three stages of decreasing pressure, especially if production is from high-pressure wells.

Initial Separation Pressure

Because of the multicomponent nature of the produced fluid, the higher the pressure at which the initial separation occurs, the more liquid that will be obtained in the separator. This liquid contains some light components that vaporize in the stock tank downstream of the separator. If the initial separation pressure is

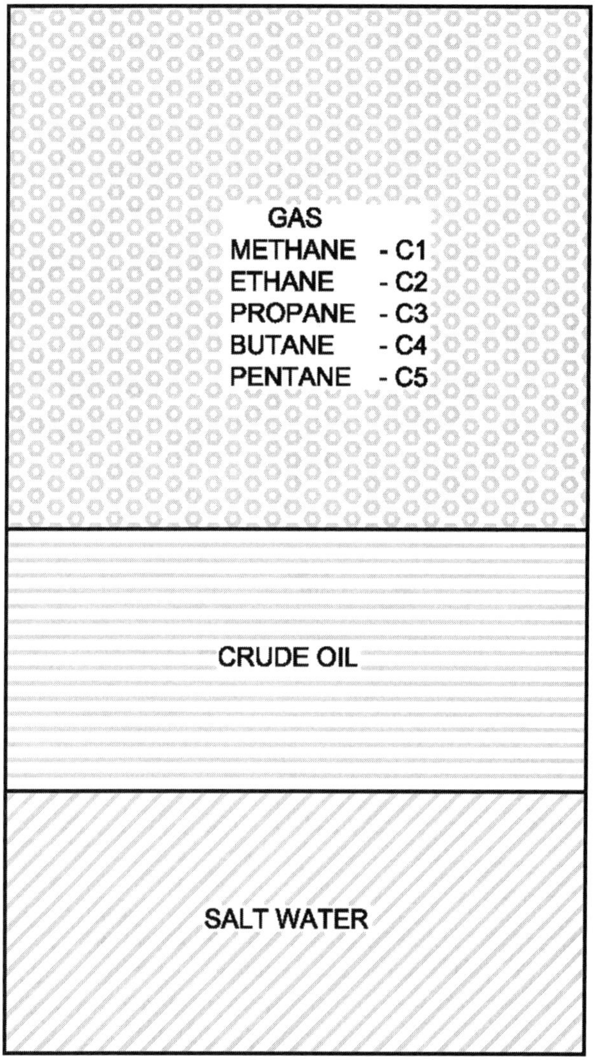

FIGURE 2.5. Typical reservoir fluids found in a well.

- too high, too many light components will stay in the liquid phase at the separator and will be lost in the tank.
- too low, not as many light components will be stabilized in the liquid phase at the separator, and they will be lost to the gas phase.

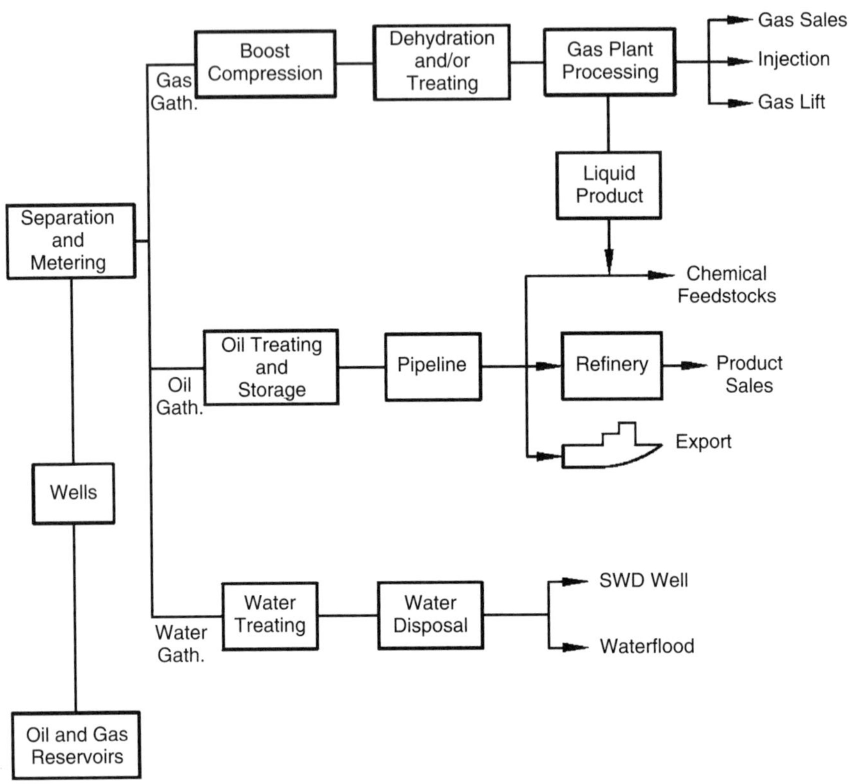

FIGURE 2.6. Major areas of activity in the production of hydrocarbons.

Single-Stage Separation

The preceding phenomenon, which can be calculated using flash calculations discussed in Chapter 1, is shown in Figures 2.7 and 2.8.

The tendency of any one component in the process stream to flash to the vapor phase depends on its partial pressure. The partial pressure of a component in a vessel is defined as the number of molecules of that component in the vapor space divided by the total number of molecules of all components in the vapor space times the pressure in the vessel. The tendency of a component to flash to gas is a function of

- pressure,
- temperature, and
- molecular composition of the fluid.

For a given temperature, this tendency can be visualized as a function of partial pressure, where

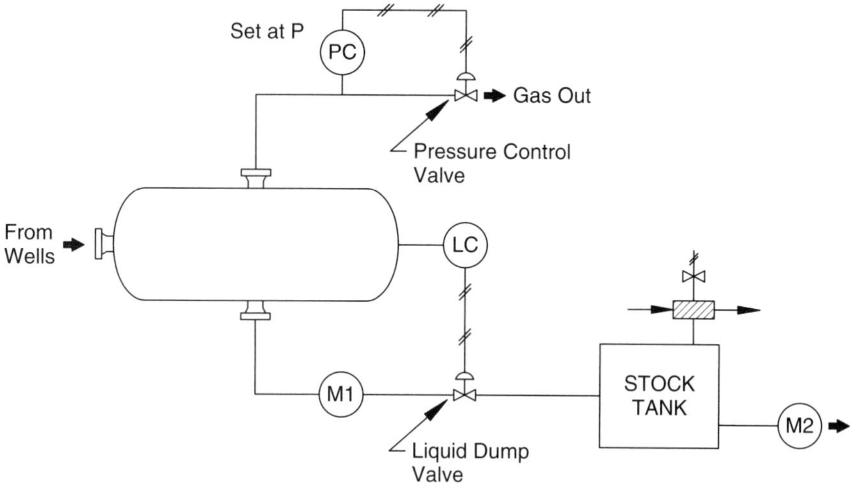

FIGURE 2.7. Single-stage separation.

$$PP_N = \frac{Moles_N}{\sum Moles_N} \text{ (Vapor pressure)} \qquad (2.1)$$

where

PP_N = partial pressure of component N,
$Moles_N$ = number of moles of component N
$\sum Moles_N$ = total number of moles of all components,
P = pressure in the vessel, psia (kPa)

The lower the partial pressure of a component, the greater the tendency that the component will flash to gas (Figure 2.7). If the pressure in the vessel is high, the partial pressure for the component will be relatively high and the molecules of that component will tend toward the liquid phase (This is seen by the top line in Figure 2.8.) As the separator pressure is increased, the liquid flow rate out of the separator increases.

The problem with this is that many of these molecules are the lighter hydrocarbons (methane, ethane, and propane), which have a strong tendency to flash to the gas state at stock-tank conditions (atmospheric pressure). In the stock tank, the presence of the large number of molecules creates a low partial pressure for the intermediate-range hydrocarbons (butane, pentane, and heptane), whose flashing tendency at stock-tank conditions is very susceptible to small changes in partial pressure. Thus, by keeping the lighter molecules in the feed to

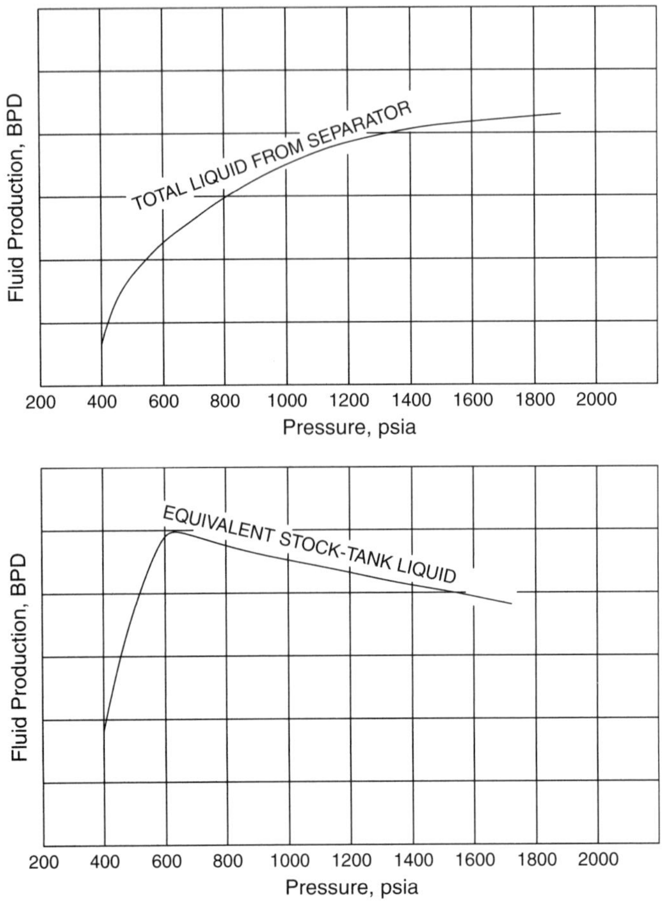

FIGURE 2.8. Effect of separator pressure on stock-tank liquid recovery.

the stock tank, we manage to capture a small amount of them as liquids, but we lose to the gas phase many more of the intermediate-range molecules. That is why beyond some optimum point there is actually a decrease in stock-tank liquids by increasing the separator operating pressure.

Stage Separation

Figure 2.7 deals with a single-stage process. Fluids are flashed in an initial separator, and then the liquids from that separator are flashed again in a stock tank. Stock tank is not normally considered a separate stage of separation, though it most assuredly is. Figure 2.9 shows a

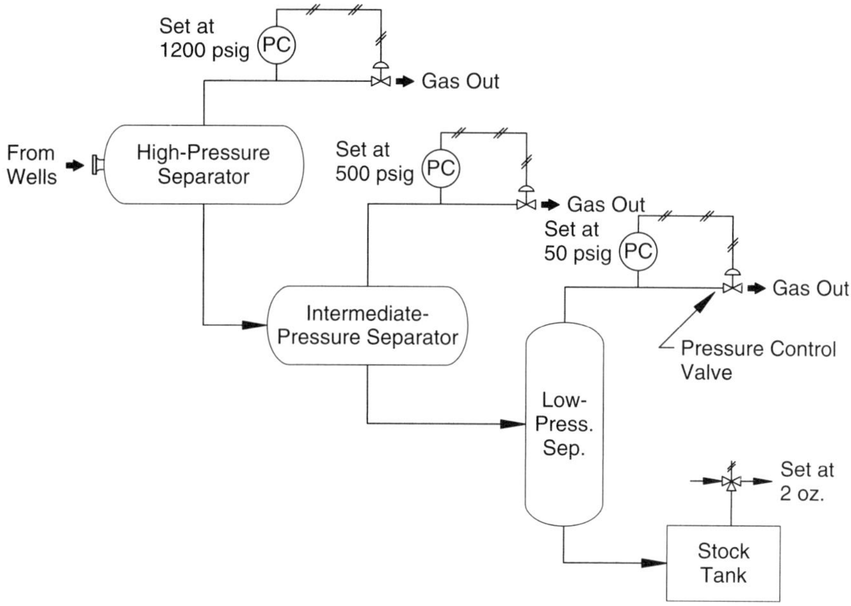

FIGURE 2.9. Stage separation.

three-stage separation process. Liquid is first flashed at an initial pressure and then flashed at successively lower pressures two times before entering the stock tank. Because of the multicomponent nature of the produced fluid, it can be shown by flash calculations that the more the stages of separation after initial separation, the more the light components will be stabilized into the liquid phase.

In a stage separation process, the light hydrocarbon molecules that flash are removed at relatively high pressure, keeping the partial pressure of the intermediate hydrocarbons lower at each stage. As the number of stages approaches infinity, the lighter molecules are removed as soon as they are formed, and the partial pressure of the intermediate components is maximized at each stage. The compressor horsepower required is also reduced by stage separation, as some of the gas is captured at a higher pressure than would otherwise have occurred (refer to Table 2.1).

Selection of Stages

As shown in Figure 2.10, as more stages are added to the process, there is less and less incremental liquid recovery. The diminishing income for adding a stage must more than offset the cost of the additional

TABLE 2.1
Effect of increasing the number of stages for a rich condensate stream

(A) Field Units			
Case	*Separation Stages (psia)*	*Liquid Produced (bopd)*	*Compressor Horsepower Required (hp)*
I	1215, 65	8400	861
II	1215, 515, 65	8496	497
III	1215, 515, 190, 65	8530	399

(B) SI Units			
Case	*Separation Stages (kPa)*	*Liquid Produced (m^3/h)*	*Compressor Horsepower Required (kW)*
I	8377, 448	55.6	642
II	8377, 3551, 448	56.3	371
III	8377, 3551, 1310, 448	56.5	298

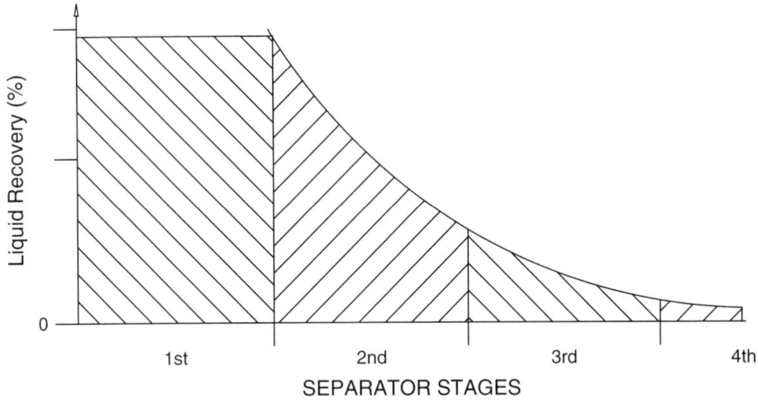

FIGURE 2.10. Incremental liquid recovery versus number of separator stages.

separator, piping, controls, space, and compressor complexities. For each facility there is an optimum number of stages. It is difficult to determine, as it may be different from well to well, and it may change as the well's flowing pressure declines with time. Table 2.2 is an approximate guide to the number of stages in separation, excluding stock tank, which field experience indicates is somewhat near

TABLE 2.2
Stage separation guidelines

Initial Separator Pressure		
Psig	*kPa*	*Number of Stages*[a]
25–125	170–860	1
125–300	860–2100	1–2
300–500	2100–3400	2
500–700	3400–4800	2–3[b]

[a]Does not include stock tank.
[b]At flow rates exceeding 100,000 BPD, stages may be appropriate.

optimum. Table 2.2 is meant as a guide and should not replace flash calculations, engineering studies, and engineering judgment.

Fields with Different Flowing Tubing Pressures

Our discussion thus far focused on a situation where all the wells in a field produce at roughly the same FTP, and stage separation is used to maximize liquid production and minimize compressor horsepower. Often, as shown in our example flowsheet (Figure 2.1), stage separation is used because different wells producing to the facility have different FTPs. This is because they are

- completed in different reservoirs or
- located in the same reservoir but have different water production rates.

Using a manifold arrangement and different separator operating pressures, provides the benefit of

- stage separation of high-pressure liquids and
- conservation of reservoir energy.

High-pressure wells can continue to flow at sales pressure requiring no compression, while wells with lower FTPs can flow into whichever system minimizes compression.

Determining Separator Operating Pressure

Choice of separator operating pressures in a multistage system is large. For large facilities handling more than 100,000 bopd, many options should be investigated before a final choice is made. For facilities handling less than 50,000 bopd, there are practical constraints that help limit the options.

Lowest-Pressure Stage

Minimum pressure is needed to move liquid through the oil and water treating systems (25–50 psig). The higher the operating pressure, the smaller the compressor needed to compress the flash gas to sales. Compressor horsepower requirements are a function of absolute discharge pressure divided by absolute suction pressure. Increasing the low-pressure separator operating pressure from 50 psig to 200 psig may decrease the required compression horsepower by 33%, but it may also add backpressure to the low-pressure wells, which

- restricts their flow and
- allows more gas flow to be vented to the atmosphere at the tank.

Usually, an operating pressure between 50 and 100 psig is optimum.

Highest-Pressure Stage

- should take advantage of reservoir energy and
- set no higher than the sales gas pressure or the required gas lift pressure, whichever is greater.

Intermediate-Pressure Stage

- useful to remember the gas from these stages must be compressed by a multistage compressor.

For practical reasons, the choice of separator operating pressures should match closely and be slightly greater than the compressor interstage pressures. The most efficient compressor sizing will be with a constant compressor ratio per stage. An approximation of the intermediate separator operating pressures can be derived from

$$R = [P_d/P_s]^{1/n} \tag{2.2}$$

where

R = ratio per stage,
P_d = discharge pressure, psia
P_s = suction pressure, psia
n = number of stages.

Once a final compressor selection is made, these approximate pressures will be changed slightly to fit the actual compressor configuration. To minimize interstage temperatures, cooling, and lubrication loads, the maximum ratio per stage is usually limited to the range of 3.6–4.0. Most facilities will have either two- or three-stage compressors. Two-stage only allows for one possible intermediate separator pressure, while a three-stage allows for either one operating at second- or third-stage suction pressure or two intermediate separators each operating at one of the two compressor intermediate suction pressures. In large facilities it is possible to install a separate compressor for each separator and operate as many intermediate-pressure separators as is deemed economical.

Two-Phase Versus Three-Phase Separators

In the example process (refer to Figure 2.1), the high- and intermediate-stage separators are two-phase, while the low-pressure separator is three-phase. The low-pressure three-phase separator is called a "free-water knockout" (FWKO) because it is designed to separate the free water from the oil and emulsion, as well as separate gas from liquid. Choice of two- or three-phase depends on the flowing characteristics of the wells.

- If large amounts of water are expected with the high-pressure wells, it is possible to reduce the size of the other separators by making the high-pressure separator three-phase.
- If individual wells are expected to flow at different FTPs, as shown in the example process (Figure 2.1), then there is no benefit of making the high-pressure separator three-phase.
- When all wells are expected to have the same FTPs at all times, it may be advantageous to remove the free water early in the separation scheme.

Process Flowsheet

Figure 2.11 is an enlargement of the FWKO shown in Figure 2.1 and shows the amount of detail expected on a flowsheet. A flash calculation is needed to determine the amount of gas and liquid that each separator must handle. In Figure 2.1, the treater is not considered a separate stage of separation as it operates very close to the FWKO pressure, which is the last stage. Very little gas will flash between the two vessels. Normally, this gas is used for fuel or vented and not compressed for sales, although a small compressor could be added to boost this gas to main compressor suction pressure.

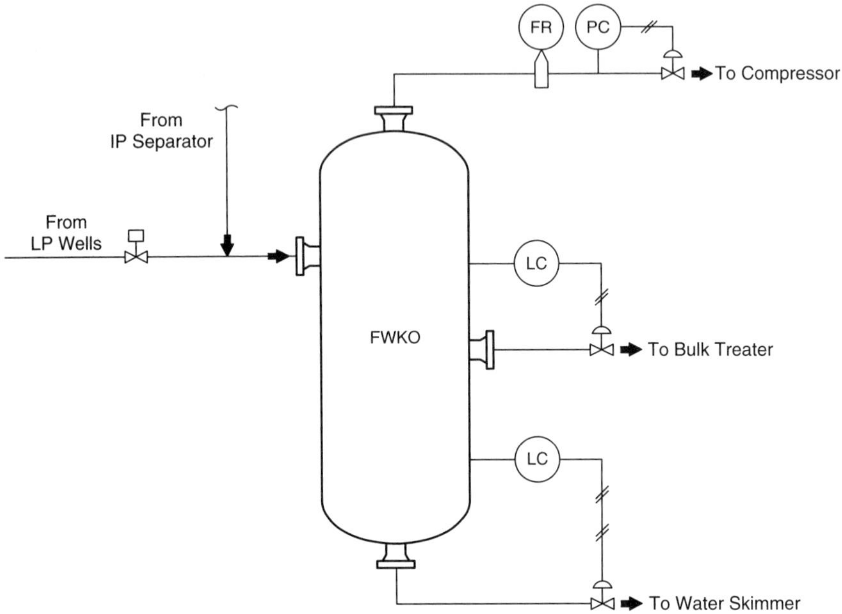

FIGURE 2.11. Vertical free-water knockout.

2.4.3 Oil Treating and Storage

Crude requires dehydration before it can go to storage. Water-in-oil emulsions must be broken so as to reduce

- water cut and
- salt content.

Demulsifier chemicals weaken the oil film around the water droplets, so the film will rupture when droplets collide. Droplet collision is accelerated by using

- heat and
- electrostatics.

Continuing surveillance is required. Treating requirements change during the depletion life of a reservoir. Revise equipment and operating procedures. Salt must also be removed from the produced crude. This is done by

- mixing fresh water with dehydrated crude and then
- dehydrating it a second time to meet TDS content requirement.

Salt content specifications range from 10 to 25 pounds per thousand barrels (PTB). Desalting is accomplished at refineries in

- USA,
- West Africa, and
- parts of southeast Asia.

Desalting is accomplished at producing fields or shipping terminals in

- Europe,
- the Middle East,
- parts of South America, and
- parts of Southeast Asia.

As the last step in production, crude may be run through a stabilizer, where its vapor pressure is reduced to allow

- nonvolatile liquid to be stored in tanks at atmospheric pressure or
- loaded onto tankers.

Offshore locations typically use vertical or horizontal treaters. Figure 2.12 is an enlargement of a horizontal oil treater in Figure 2.1.

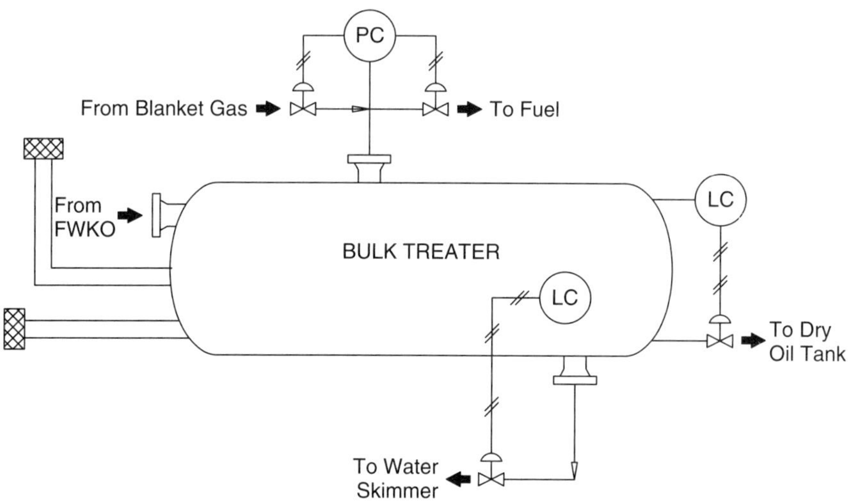

FIGURE 2.12. Horizontal bulk treater.

Gas blanket is provided to

- ensure that there is always sufficient pressure in the treater to allow the water to flow to the water treating system without requiring a pump and
- excludes oxygen entry, which could cause scale, corrosion, and bacteria.

Onshore locations typically use a "Gunbarrel" (wash tank/settling tank) with either an internal or external "Gas Boot." Figure 2.13 is an enlargement of a Gunbarrel with an internal Gas Boot. A Gunbarrel with internal gas boot is used for low to moderate flow rates (1500–3000 bopd). Gunbarrel (wash tank) with external gas boot is used in low-pressure, large flow-rate systems (5000+ bopd).

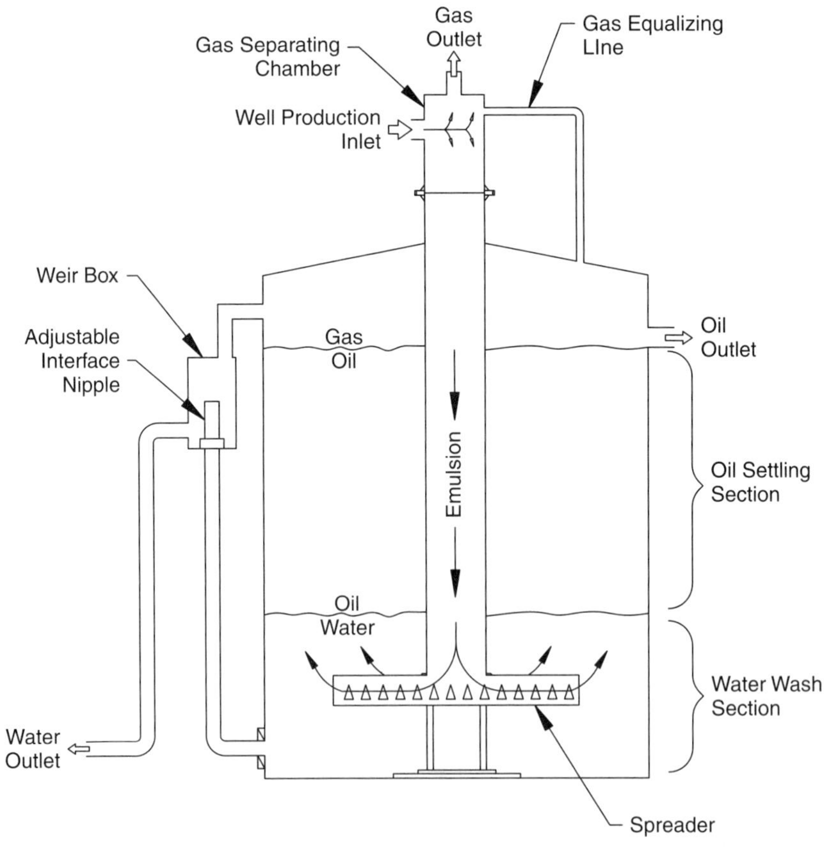

FIGURE 2.13. Gunbarrel with an internal Gas Boot.

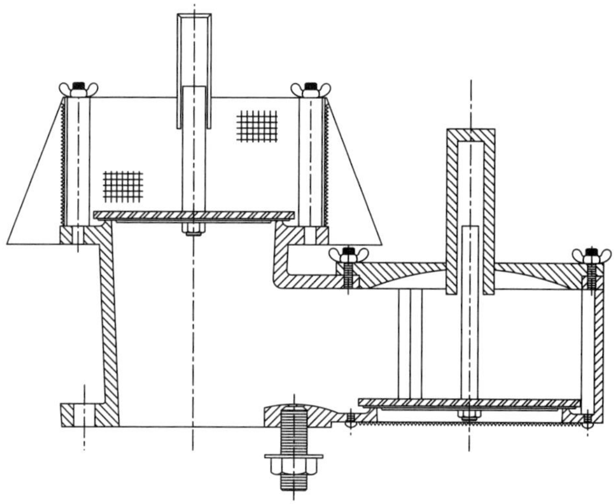

FIGURE 2.14. Typical pressure/vacuum valve (courtesy of Groth Equipment Corp.).

All tanks should have a pressure/vacuum valve with a flame arrestor and a gas blanket to keep a positive pressure on the system and exclude oxygen.

- Figure 2.14 is an enlargement of a typical pressure/vacuum valve.
- Figure 2.15 is an enlargement of a typical flame arrestor.
- Table 2.3 shows the savings associated with keeping a positive pressure on a tank.

Oil is skimmed off the surface of the Gunbarrel or wash tank, and the water exits from the bottom through either a water leg or an interface level controller and dump valve. Since the volume of the liquid is fixed by the oil outlet, Gunbarrels and wash tanks cannot be used as surge tanks. Flow from the treater or Gunbarrel goes to a settling/shipping tank, from which it either flows into a barge or truck or is pumped into a pipeline.

2.4.4 Lease Automatic Custody Transfer (LACT)

Large facilities usually sell oil through a LACT unit. LACT units are designed to meet API Standards and whatever additional measuring and sampling standards are required by the crude purchaser.

Value received for the crude depends on

- gravity,
- basic settlement and water (BS&W) content, and
- volume.

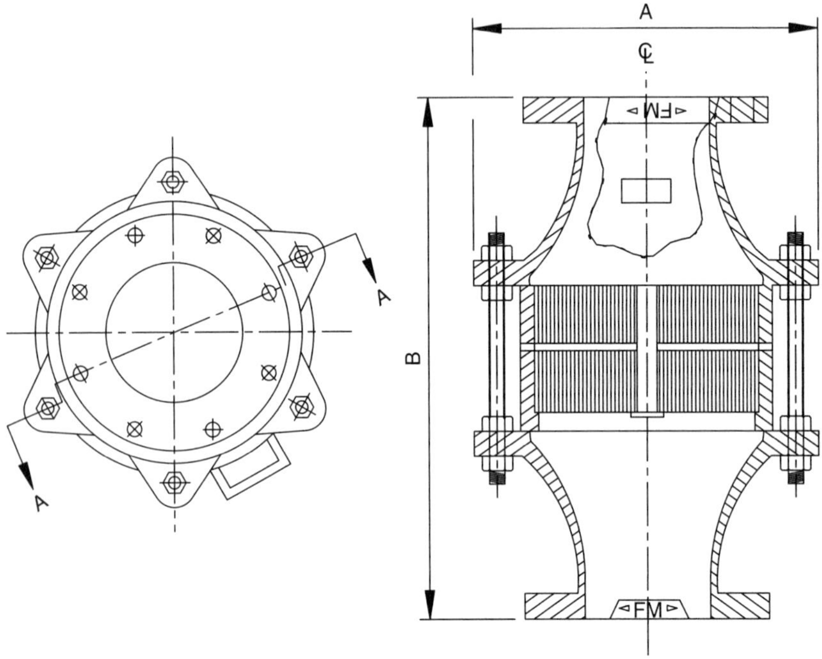

FIGURE 2.15. Typical frame arrestor (courtesy of Groth Equipment Corp.).

TABLE 2.3
Tank breathing loss

Nominal Capacity (BBLS)	*Breathing Loss* *Open Vent (BBL/yr)*	*Pressure Valve (BBL/yr)*	*Barrels Save*
5000	235	154	81
10,000	441	297	144
20,000	625	570	255
55,000	2000	1382	618

Figure 2.16 shows a schematic of the elements of a typical LACT unit. Crude first flows through a strainer/gas eliminator to protect the meter and to ensure that there is no gas in the liquid. When BS&W exceeds the sales contract quality, this probe automatically actuates the diverter valve, which blocks the liquid from going further in the LACT unit and sends it back to the process for further treating. Some sales contracts allow for the BS&W probe to merely sound a warning so that the operators can manually take corrective action. In this

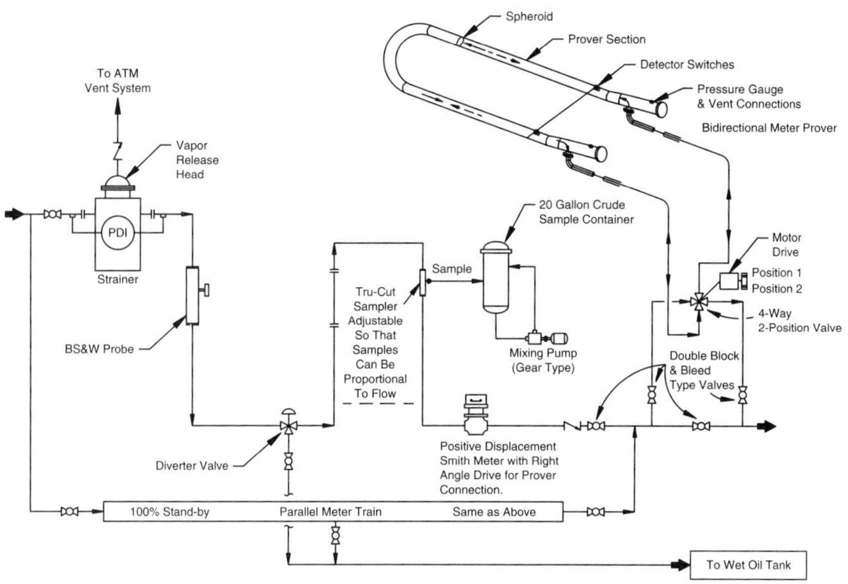

FIGURE 2.16. Typical LACT unit schematic.

situation, the unit is called an ACT and not a LACT. The BS&W probe must be mounted in a vertical run if it is to get a true reading of the average quality of the stream. Downstream of the diverter valve is a sampler located in the vertical run. Sampler takes a calibrated sample that is proportional to the flow and delivers it to a sample container. The sampler receives a signal from the meter to ensure that the sample size is always proportional to flow even if the flow varies. Sample container has a mixing pump so that the liquid in the container can be mixed and made homogeneous prior to taking a sample of this fluid. Sample contained in the sample container is used to convert the meter reading for BS&W and gravity. Liquid then flows through a positive displacement meter. Most sales contracts require the meter to be proven at least once a month and a new meter factor calculated.

On large installations, a meter prover such as that shown in Figure 2.16 is included as a permanent part of the LACT skid or is brought to the location when a meter must be proven. Meter prover contains a known volume between two detector switches. Volume recorded by the meter during the time the psig moves between detectors for a set number of traverses of the prover is recorded electrically and compared to the known volume of the meter prover. On smaller installations, a master meter that has been calibrated using a calibrated prover may be brought to the location to run in series with the meter to be proven.

2.4.5 Pumps

Pumps are normally needed to

- move oil through the LACT unit and
- deliver oil to a pipeline downstream of the LACT unit.

Pumps are sometimes used in water-treating and disposal processes. Small pumps may be required to pump skimmed oil to higher-pressure vessels for treating glycol heat medium, cooling water service, firefighting, and so forth.

2.4.6 Water Treating

Figure 2.17 shows an enlargement of the water-treating system as an example process flowsheet.

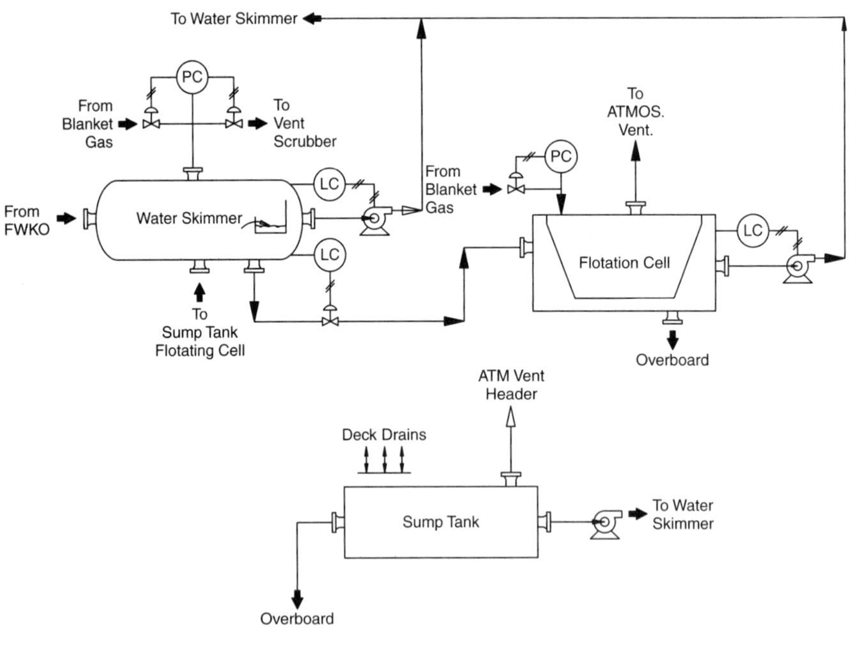

FIGURE 2.17. Water treating system.

2.4.7 Compressors

Figure 2.18 shows the configuration of a typical three-stage reciprocating compressor in our example flowsheet. Gas from the FWKO enters the first-stage suction scrubber. Any liquids that may have come through the line are separated at this point and the gas flows to the first stage.

Compression heats the gas, so there is a cooler after each compression stage. At the higher pressure, more liquids may separate, so the gas enters another scrubber before being compressed and cooled again.

In the example flowsheet, gas from the intermediate-pressure separator can be routed to either the second-stage or third-stage suction pressure, as conditions in the field change.

Reciprocating compressors are attractive for

- low horsepower (<2500 hp) and
- high-ratio applications (5–20)

Reciprocating compressors have

- higher efficiencies than centrifugals and
- much higher turndown capabilities.

Centrifugal compressors are attractive for

- high horsepower (>4000 hp) and
- low-ratio applications (2–5).

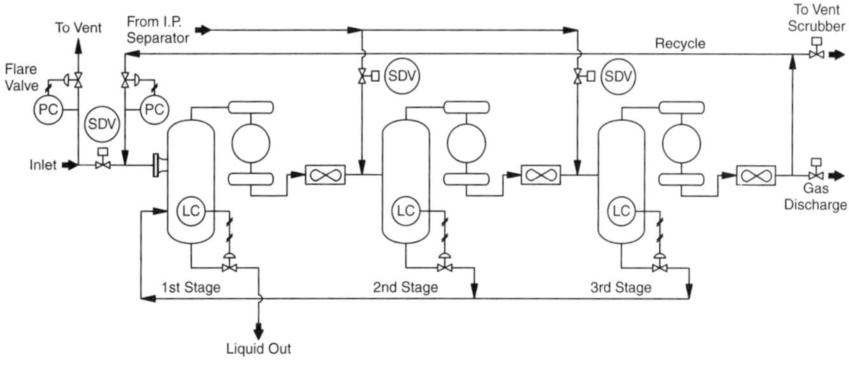

FIGURE 2.18. Three-stage compressor.

Centrifugal compressors

- are less expensive,
- take up less space,
- weigh less, and
- tend to have higher availability and lower maintenance costs.

2.4.8 Gas Dehydration

Removing most of the water vapor from the gas is required by most gas sales contracts, because it

- prevents hydrates from forming when the gas is cooled in the transmission and distribution systems and
- prevents water vapor from condensing and creating a corrosion problem.

Dehydration also marginally increases line capacity. Most sales contracts call for reducing the water content in the gas to less than 7 lb/MMscf. In colder climates, sales requirements of 3–5 lb/MMscf are common.

The following methods can be used for drying the gas:

- Cool to the hydrate formation level and separate the water that forms. This can only be done where high water contents ($\pm$30 lb/MMscfd) are acceptable.
- Use a low-temperature exchange (LTX) unit designed to melt the hydrates as they are formed. Figure 2.19 shows the process. LTX units require inlet pressures greater than 2500 psi to work effectively. Although they were common in the past, they are not normally used because of their tendency to freeze and their inability to operate at lower pressures as the well FTP declines.
- Contact the gas with a solid bed of $CaCl_2$. The $CaCl_2$ will reduce the moisture to low levels, but it cannot be regenerated and is very corrosive.
- Use a solid desiccant, such as activated alumina, silica gel, or molecular sieve, which can be regenerated. These are relatively expensive units, but they can get the moisture content to very low levels. Therefore, they tend to be used on the inlets to low-temperature gas processing plants but are not common in production facilities.
- Use a liquid desiccant, such as methanol or ethylene glycol, which cannot be regenerated. These are relatively inexpensive.

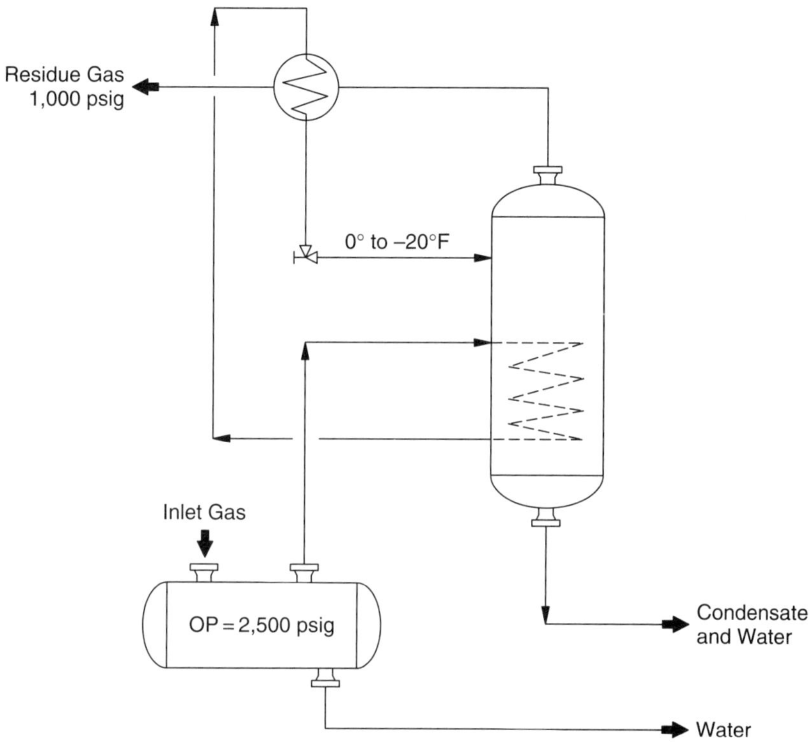

FIGURE 2.19. Low-temperature exchange unit.

Extensive use is made of methanol to lower the hydrate temperature of gas well flowlines to keep hydrates from freezing the choke.

- Use a glycol liquid desiccant, which can be regenerated. This is the most common type of gas dehydration system and is the one shown in the example process flowsheet.

Figure 2.20 shows how a typical bubble-cap glycol contact tower works. Wet gas enters the base of the tower and flows upward through the bubble caps. Dry glycol enters the top of the tower, because of the down-comer weir on the edge of the tray, flows across the tray, and down to the next. There are typically six to eight trays in most applications. The bubble caps ensure that the upward-flowing gas is dispersed into small bubbles to maximize its contact area with the glycol.

Before entering the contactor, the dry glycol is cooled by the outlet gas to condense water vapor and hydrocarbon liquids as much as

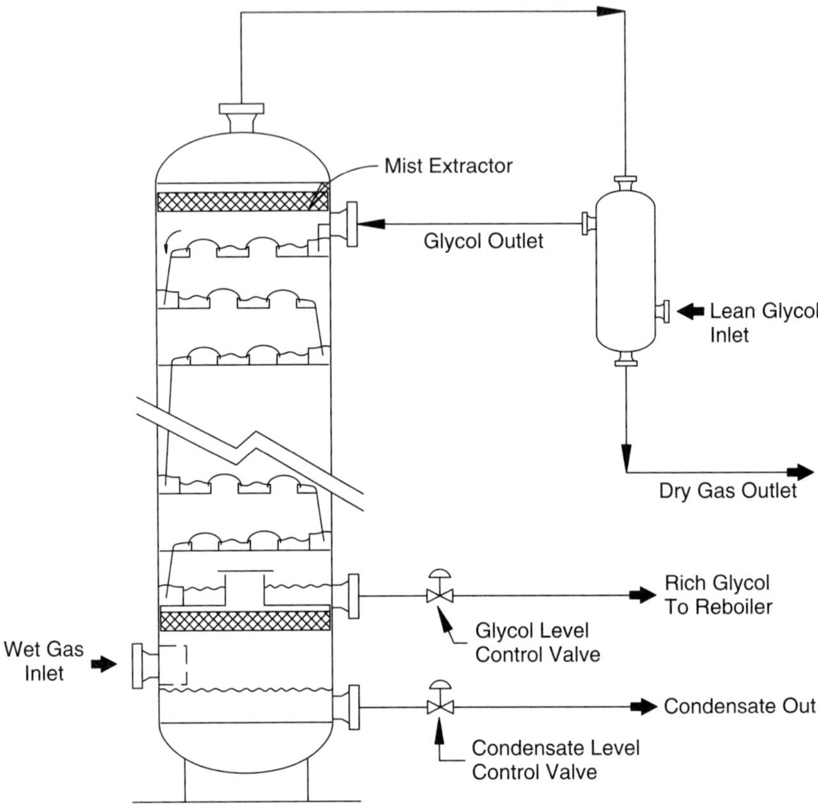

FIGURE 2.20. Typical glycol contact tower.

possible before it enters the tower. The wet glycol leaves from the base of the tower and flows to the reconcentrator (reboiler) by way of heat exchangers, a gas separator, and filters, as shown in Figure 2.21. In the reboiler, the glycol is heated to a sufficiently high temperature to drive off the water as steam. The dry glycol is then pumped back to the contact tower.

Most glycol dehydrators use triethylene glycol, which can be heated to 340–400°F in the reconcentrator and work with gas temperatures up to 120°F. Tetraethylene glycol is more expensive, but it can handle hotter gas without experiencing high glycol losses and can be heated in the reconcentrator to 400–430°F.

2.5 Well Testing

Well testing allows one to keep track of the oil, gas, and water production from each well so as to

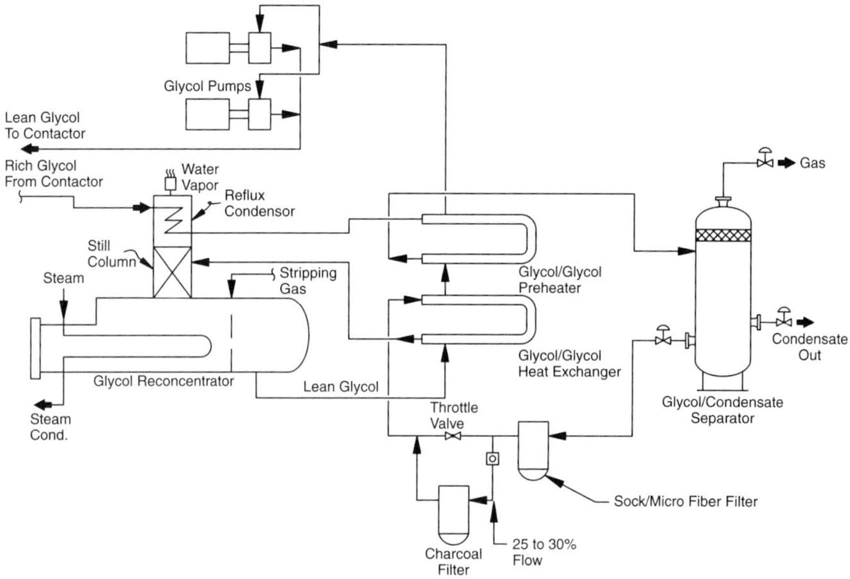

FIGURE 2.21. Typical glycol reconcentrator.

- manage the reserves properly,
- evaluate where further reserve potential may be found, and
- diagnose well problems as quickly as possible.

Proper allocation of income also requires knowledge of daily production rates as the royalty or working interest ownership may be different for each well.

In simple facilities that contain only a few wells, it is attractive to route each well to its own separator and/or treater and measure its gas, oil, and water production on a continuous basis.

In facilities that handle production from many wells, it is sometimes more convenient to enable each well to flow through the manifold to one or more test subsystems.

Some facilities use a high-pressure three-phase separator for the high- and intermediate-pressure wells that do not make much water and a treater for the low-pressure wells. Figure 2.22 shows an enlargement of the well test separator.

2.6 Gas Lift

Figure 2.23 is a diagram of a gas lift system from the facility engineer's perspective.

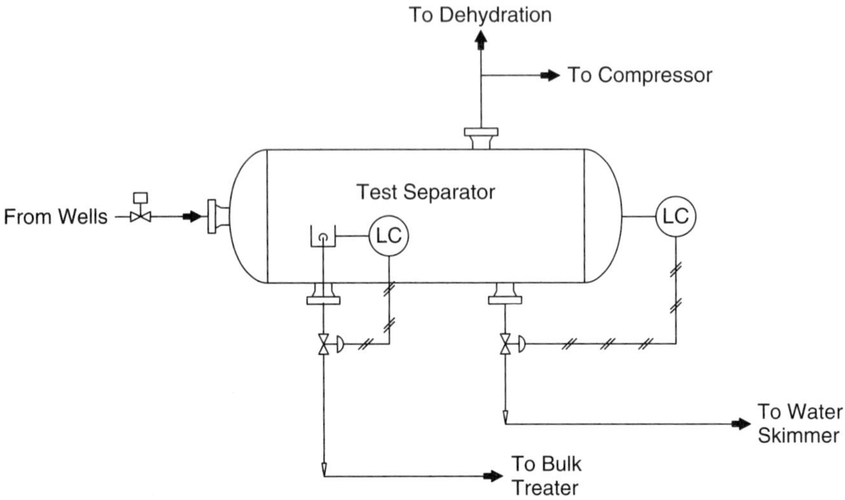

FIGURE 2.22. Well test system.

High-pressure gas is injected into the well to lighten the column of fluid and allow the reservoir pressure to force the fluid to the surface.

The gas that is injected is produced with the reservoir fluid into the low-pressure system. The low-pressure separator must have sufficient gas separation capacity to handle gas lift as well as formation gas.

Figure 2.24 shows the effects of wellhead backpressure for a specific set of wells. A 1-psi change in well backpressure will cause between a 2- and 6-BFPD change in well deliverability. If gas lift is to

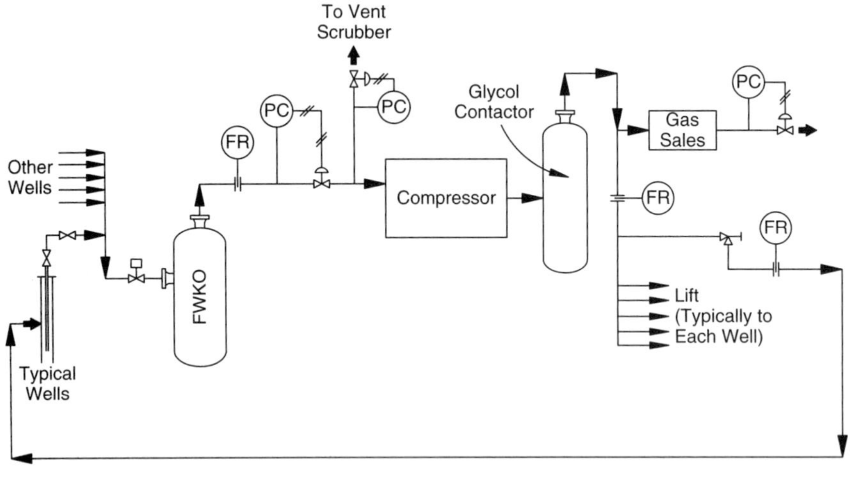

FIGURE 2.23. Gas lift system.

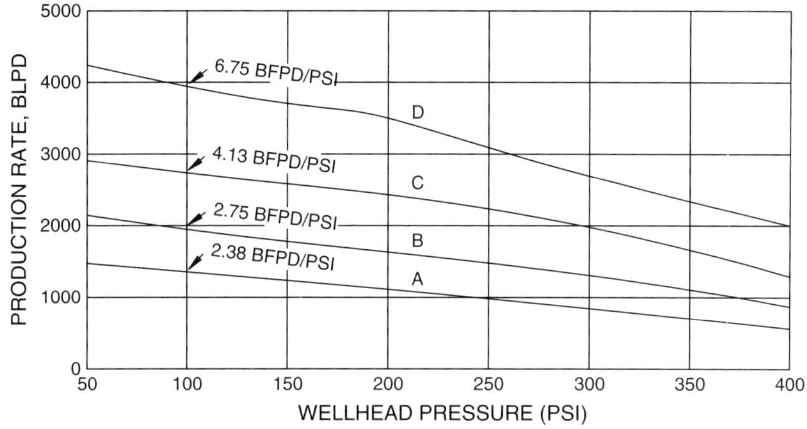

Note: These curves are for a specific set of tubing size, casing pressure, and fluid out.

FIGURE 2.24. Effect of wellhead backpressure on total fluid production rate for a specific set of wells.

be used, it is even more important from a production standpoint that the low-pressure separator be operated at the lowest practical pressure.

Figure 2.25 shows that for a typical well, the higher the design injection, the higher the flow rate. The higher the injected gas pressure into the casing, the deeper the last gas lift valve can be set.

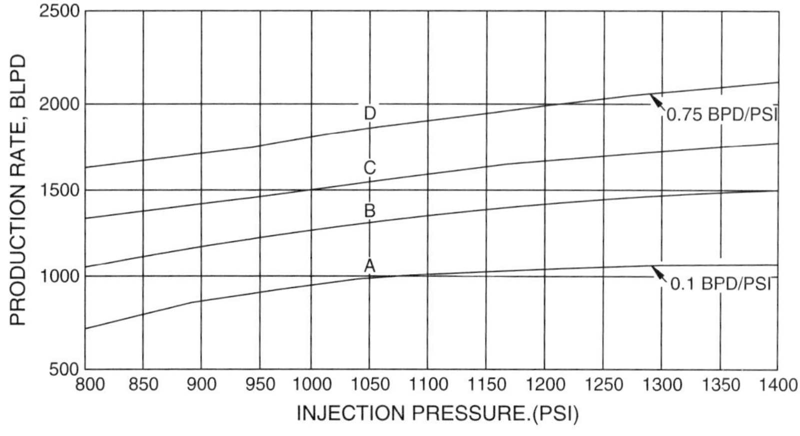

Note: These curves are for a specific set of tubing size, casing pressure, and fluid out.

FIGURE 2.25. Effect of gas lift injection pressure on total fluid production rate for a specific set of wells.

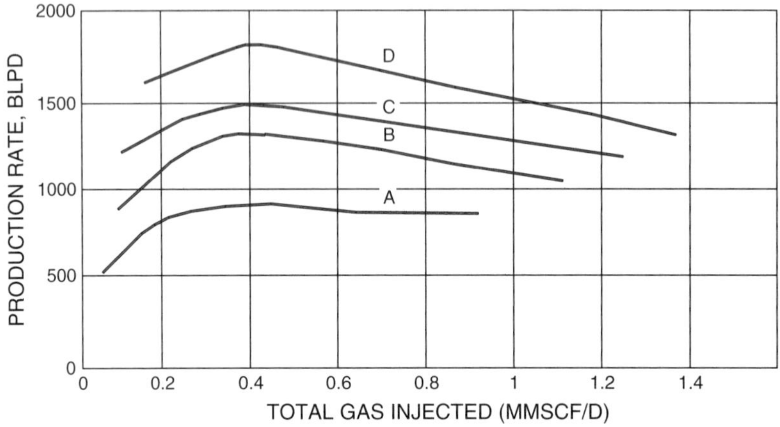

FIGURE 2.26. Effect of gas lift injection rate on total fluid production rate for a specific set of wells.

Figure 2.26 shows the effect of gas injection rate. As more gas is injected, the weight of fluid in the tubing decreases and the bottom-hole flowing pressure decreases.

2.7 Offshore Platform Considerations

2.7.1 Overview

An increasing amount of the world's oil and gas comes from offshore fields. This section describes platforms that accommodate simultaneous drilling and production operations.

2.7.2 Modular Construction

Modules are large boxes of equipment installed in place and weighing from 300 to 2000 tons each. Modules are constructed, piped, wired, and tested in shipyards or in fabrication yards and transported on barges and set on the platform, where the interconnections are made (Figure 2.27). Modular construction is used to reduce the amount of work and the number of people required for installation and start-up.

2.7.3 Equipment Arrangement

The equipment arrangement plan shows the layout of all major equipment. Each platform has a unique layout requirement based on drilling and well-completion needs that differ from installation to installation. Layouts can be on one level or multiple levels. An example layout is shown in Figure 2.28.

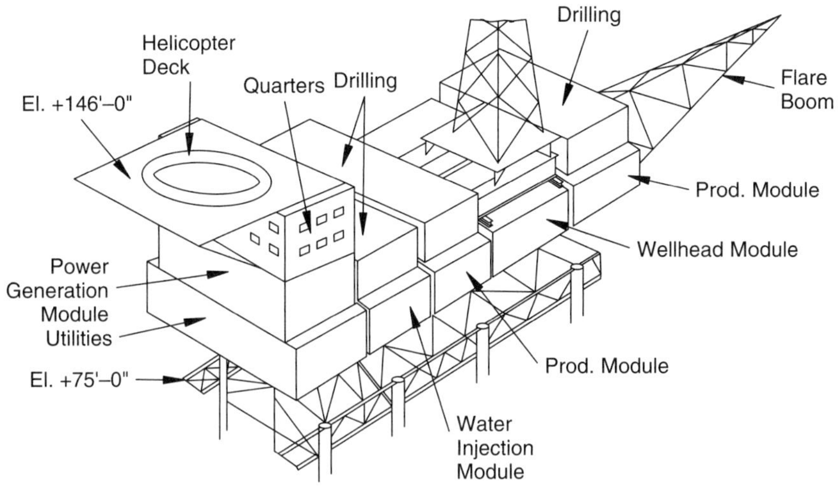

FIGURE 2.27. Schematic of a large offshore platform, illustrating the concept of modularization.

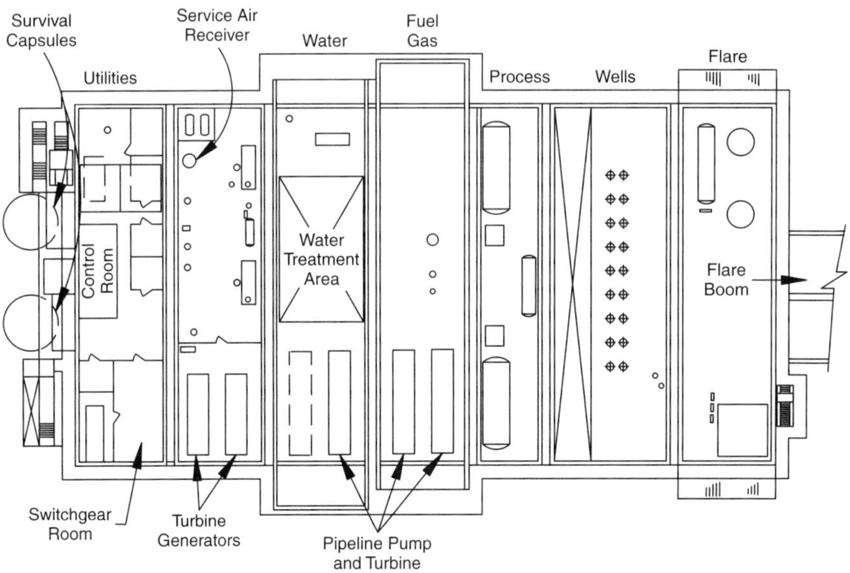

FIGURE 2.28. Equipment arrangement plan of a typical offshore platform illustrating the layout of the lower deck.

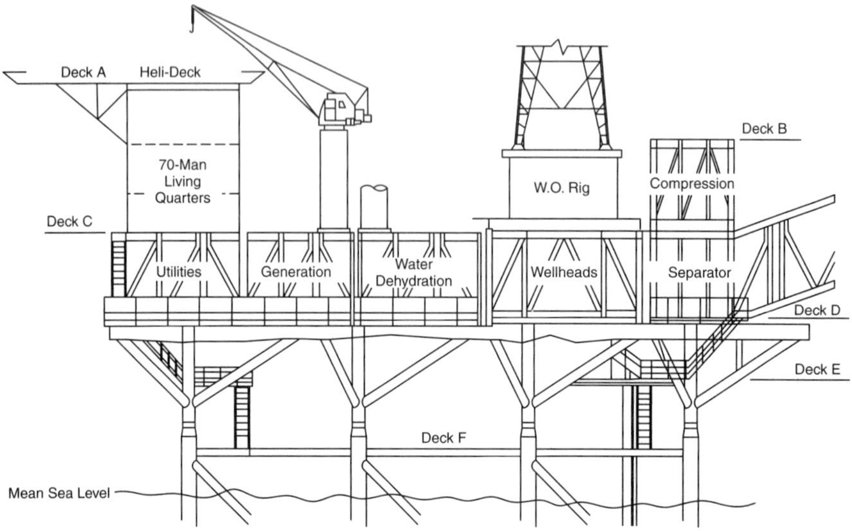

FIGURE 2.29. Typical elevation view of an offshore platform showing the relationship among the major equipment modules.

The right-hand module contains the flare drums, water skimmer tank, and some storage vessels. It also provides support for the flare boom. The adjacent wellhead module consists of a drilling template with conductors through which the wells will be drilled. The third unit from the right contains the process module, which houses the separators and other processing equipment. The fourth and fifth modules house utilities such as power generators, air compressors, potable water makers, a control room, and switchgear and battery rooms. The living quarters are located over the last module. Figure 2.29 shows an elevation of a platform in which the equipment arrangement is essentially the same.

CHAPTER 3

Two-Phase Gas–Liquid Separators

3.1 Introduction

In oil and gas separator design, we mechanically separate from a hydrocarbon stream the liquid and gas components that exist at a specific temperature and pressure.

Proper separator design is important because a separation vessel is normally the initial processing vessel in any facility, and improper design of this process component can "bottleneck" and reduce the capacity of the entire facility.

A separator is a pressure vessel designed to divide a combined liquid–gas system into individual components that are relatively free of each other for subsequent disposition or processing.

Downstream equipment cannot handle gas–liquid mixtures, for example:

- Pumps require gas-free liquid;
- Compressor and dehydration equipment require liquid-free gas;
- Product specification set limits on impurities
  - Oil generally cannot contain more than 1% BS&W
  - Gas sales contracts generally require that the gas contain no free liquids; and
- Measurement devices for gases or liquids are highly inaccurate when another phase is present.

Separators are sometimes called "gas scrubbers" when the ratio of gas rate to liquid rate is very high. A "slug catcher," commonly used in gas gathering pipelines, is a special case of a two-phase gas–liquid separator that is designed to handle large gas capacities and liquid slugs.

3.1.1 Characteristics of the Flow Stream

Fluid from a well can include:

- gas
- condensable liquid vapors
- water
- water vapor
- crude oil
- solid debris

The proportion of each of the above components varies from well to well. Well fluids exist as either

- emulsion (Figure 3.1)
- layered (Figure 3.2)

Free fluids separate more easily than fluids in an emulsion. Solution gas is gas dissolved in well fluids, rather than carried in the stream. Solution gas is not free. As the pressure on well fluids decreases, the capacity of liquid to hold gas in solution decreases. As well fluids

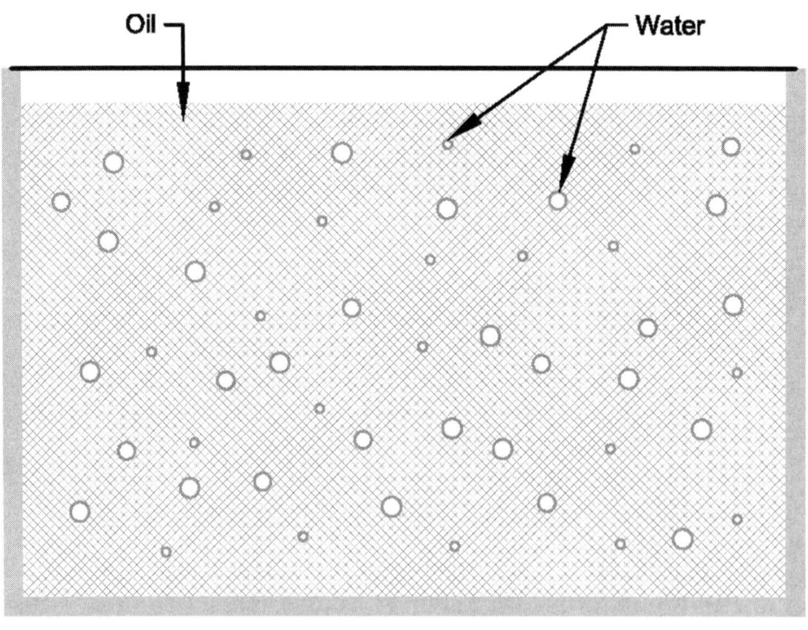

FIGURE 3.1. Emulsion where oil is mixed with small droplets of water that are coated with oil.

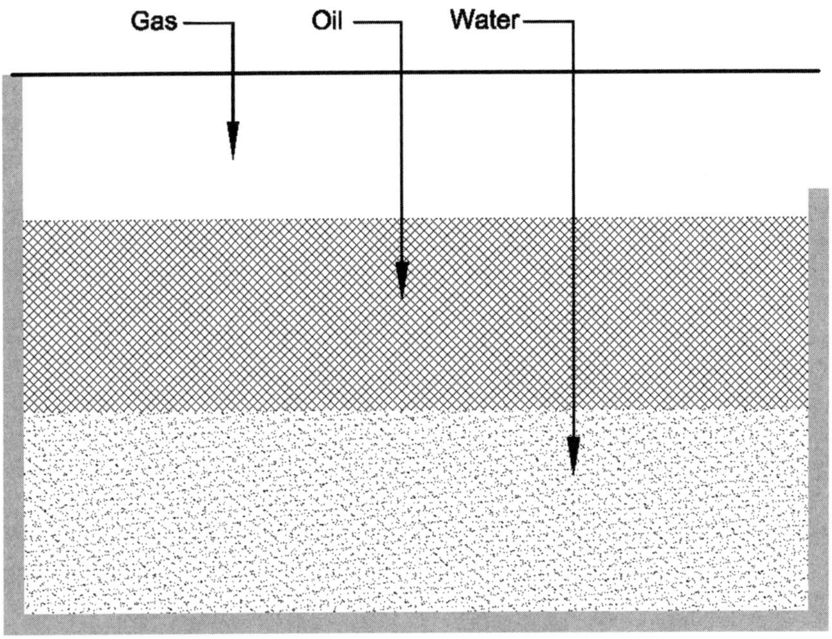

FIGURE 3.2. Layered fluids.

reach ground level, the capacity of liquid to hold solution gas decreases and the gas separates out of the oil.

Wells are classified according to the type of fluid they produce in the greatest quantity.

- Crude oil well
  - contains mostly crude oil, but can contain
    - solid debris
    - water
    - gas
- Dry gas well
  - contains mostly gas
  - can contain some water
  - does not contain crude or liquid hydrocarbons
- Gas condensate well
  - contains both liquid and gaseous hydrocarbons
  - contains some water
  - does not contain crude oil

A condensate hydrocarbon is a very light hydrocarbon that changes from liquid to vapor at near atmospheric conditions. Gas that is produced with oil is called casing head gas or associated gas, while gas

TABLE 3.1
Well classifications, fluid components and processing

Class of Well	Fluids in the Reservoir	Fluids in Flow Line	Processing Step That May Be Required
Dry gas	Gas, possibly water	Gas, possibly water	Separation, gas dehydration
Gas condensate	Gas, possibly water	Gas, condensate, possibly water	Separation, gas dehydration, condensate stabilization
Crude oil	Crude oil possibly gas possibly water	Crude oil, possibly gas, possibly water	Separation, gas dehydration, crude stabilization

produced alone or with water is called non-associated gas. Table 3.1 is a summary of well classifications, fluid components, and processing methods.

3.1.2 Well Fluids

Reservoir pressures are generally much higher than atmospheric pressure. As well fluids reach the surface, the pressure on them is decreased and the ability to hold gas in solution decreases. Solution gas released as free gas is held by the surface tension of the oil. Surface tension is reduced when the well fluids are warmed. Gravity alone will cause the heavy components to settle out and the light components to rise. Three variables that aid in separation are temperature, pressure, and density.

Well fluid separation depends on the composition of the fluids and the pressure and temperature. Pressure on the fluids is controlled by a back pressure control valve. Temperature of the fluids is regulated by expanding the fluids through a choke, heating the fluids in a heater treater, and heating or cooling the fluids in a heat exchanger.

Separators can be designed to handle fluids according to the fluid composition. Gas–liquid separators (two-phase) separate well fluid into its liquid and gaseous components. Liquid–liquid separators (three-phase) separate well fluid into water, oil, and gas.

3.1.3 Phase Equilibrium

The phase equilibrium diagram is a useful tool to visualize phase behavior. Phase equilibrium is a theoretical condition where the

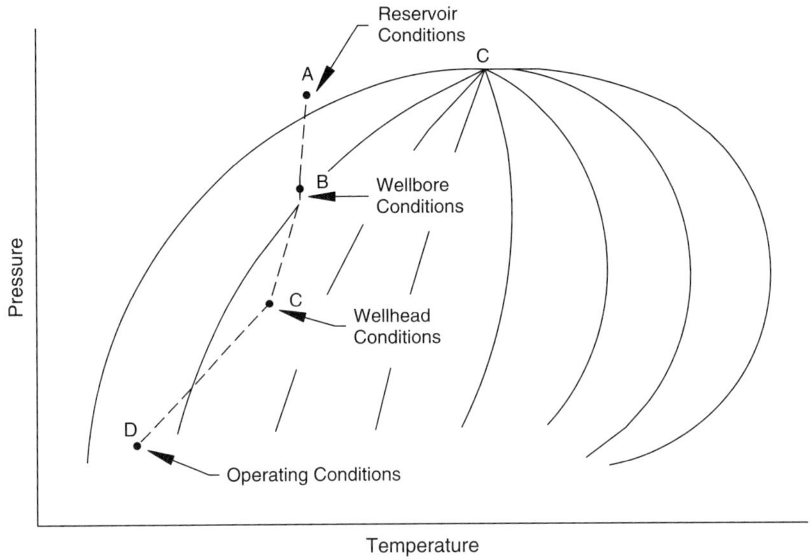

FIGURE 3.3. Phase equilibrium phase diagram for a typical production system.

liquids and vapors have reached certain pressure and temperature conditions at which they can separate. Figure 3.3 illustrates several operating points on a generic phase equilibrium diagram.

- Point A represents the operating pressure and temperature in the petroleum reservoir (liquid).
- Point B represents the flowing conditions at the bottom of the production tubing of a well (two-phase).
- Point C represents the flowing conditions at the wellhead. Typically, these conditions are called flowing tubing pressure (FTP) and flowing tubing temperature (FTT).
- Point D represents the surface conditions at the inlet of the first separator (two-phase).

3.1.4 Factors Affecting Separation

Characteristics of the flow stream will greatly affect the design and operation of a separator. The following factors must be determined before separator design:

- gas and liquid flow rates (minimum, average, and peak),
- operating and design pressures and temperatures,

- surging or slugging tendencies of the feed streams,
- physical properties of the fluids such as density and compressibility factor,
- designed degree of separation (e.g., removing 100% of particles greater than 10 μm),
- presence of impurities (paraffin, sand, scale, etc.),
- foaming tendencies of the crude oil, and
- corrosive tendencies of the liquids or gas.

3.2 Functional Sections of a Gas–Liquid Separator

3.2.1 Introduction

The separator sections described below utilize gravity settling, velocity separation by centrifugal force or impingement, and filtration. Additional methods of separation are sometimes required after primary separation, such as thermal (crude oil heater-treaters), electrostatic precipitation (crude oil electrostatic coalescing treaters), adhesive separation (gas-filter separators and water clean-up precipitators), and adsorption (gas molecular sieves, silica gels, and alumina gels).

Regardless of the size or shape of a separator, each gas–liquid separator contains four major sections. Figures 3.4 and 3.5 illustrate the four major sections of a horizontal and vertical two-phase gas–liquid separator.

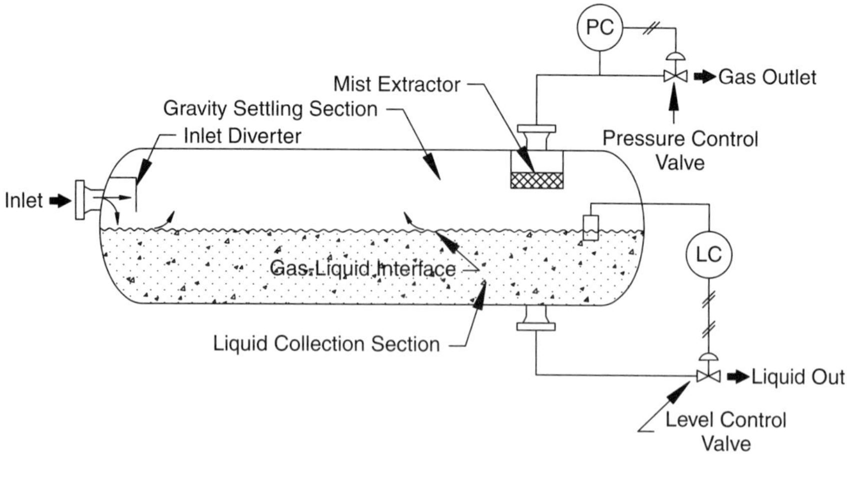

FIGURE 3.4. Horizontal separator schematic.

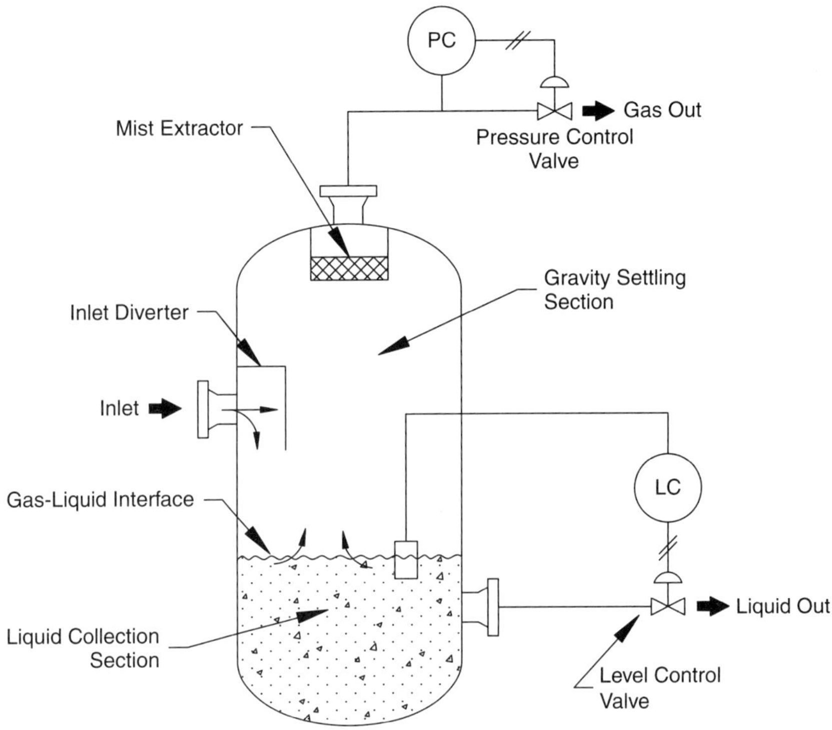

FIGURE 3.5. Vertical separator schematic.

3.2.2 Inlet Diverter

This abruptly changes the direction of flow by absorbing the momentum of the liquid and gas to separate. This results in the initial "gross" separation of liquid and gas.

3.2.3 Gravity Settling Section

This section is sized so that liquid droplets greater than 100–140 μm fall to the gas–liquid interface, while smaller liquid droplets remain with the gas. Liquid droplets, greater than 100 μm, are undesirable as they can overload the mist extractor at the separator outlet.

3.2.4 Mist Extractor Section

Before the gas leaves the vessel, it passes through a coalescing section or mist extractor. This section uses coalescing elements that provide a large amount of surface area used to coalesce and remove the small droplets of liquid. As the gas flows through the coalescing elements, it must make numerous directional changes. Due to their greater mass, the liquid droplets cannot follow the rapid changes in direction of flow. These droplets impinge and collect on the coalescing elements, where they fall to the liquid collection section.

3.3 Equipment Description

Separators are designed and manufactured in horizontal, vertical, spherical, and a variety of other configurations. Each configuration has specific advantages and limitations. Selection is based on obtaining the desired results at the lowest "life-cycle" cost.

3.3.1 Horizontal Separators

Figure 3.6 is a cutaway of a horizontal two-phase separator. Fluid enters the separator and hits an inlet diverter, causing a sudden change in momentum.

The initial gross separation of liquid and vapor occurs at the inlet diameter. The force of gravity causes the liquid droplets to fall out of the gas stream to the bottom of the vessel, where it is collected.

The liquid collection section provides

- the retention time required to let entrained gas evolve out of the oil and rise to the vapor space and reach a state of equilibrium, and
- a surge volume, if necessary, to handle intermittent slugs of liquid.

The liquid leaves the vessel through the liquid dump valve. The liquid dump valve is regulated by a level controller. The level

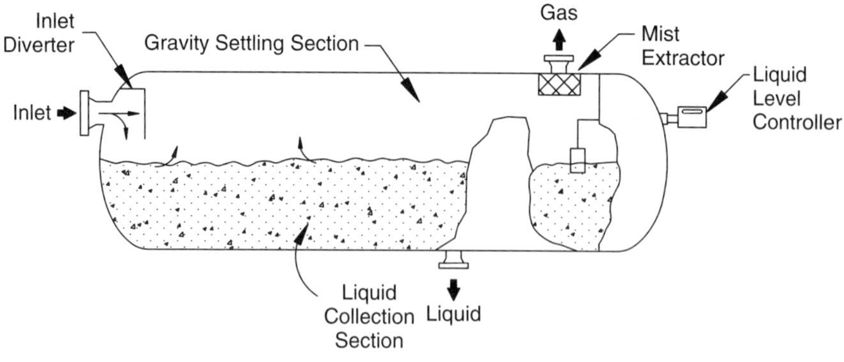

FIGURE 3.6. Cutaway view of a horizontal two-phase separator.

controller senses changes in liquid level and controls the dump valve accordingly.

Gas and oil mist flow over the inlet diverter and then horizontally through the gravity settling section above the liquid. As the gas flows through this section, small droplets of liquid that were entrained in the gas and not separated by the inlet diverter are separated out by gravity and fall to the gas–liquid interface.

Some of the drops are of such a small diameter that they are not easily separated in the gravity settling section. Before the gas leaves the vessel, it passes through a coalescing section or mist extractor that removes very small droplets of liquid in one final separation before the gas leaves the vessel.

The pressure in the separator is maintained by a pressure controller mounted on the gas outlet.

Horizontal separators are

- smaller and thus less expensive than a vertical separator for a given gas and liquid flow rate, and
- commonly used in flow streams with high gas–liquid ratios and foaming crude.

3.3.2 Vertical Separators

Figure 3.7 is a cutaway of a vertical two-phase separator. Inlet flow enters the vessel through the side. The inlet diverter does the initial gross separation. The liquid flows down to the liquid collection section of the vessel. There are seldom any internals in the liquid collection section except possibly a still well for the level control float or displacer.

Liquid continues to flow downward through this section to the liquid outlet. As the liquid reaches equilibrium, gas bubbles flow counter to the direction of the liquid flow and eventually migrate to the vapor space.

The level controller and the dump valve operate the same as in a horizontal separator. The gas flows over the inlet diverter and then vertically upward toward the gas outlet.

Secondary separation occurs in the upper gravity settling section. Liquid droplets fall vertically downward counter-current to the upward gas flow. The settling velocity of a liquid droplet is directly proportional to its diameter. If the size of the liquid droplet is too small, it will be carried up and out with the vapor.

A mist extractor section is added to capture small liquid droplets. Gas goes through the mist extractor section before it leaves the vessel. Pressure and level are maintained as in a horizontal separator.

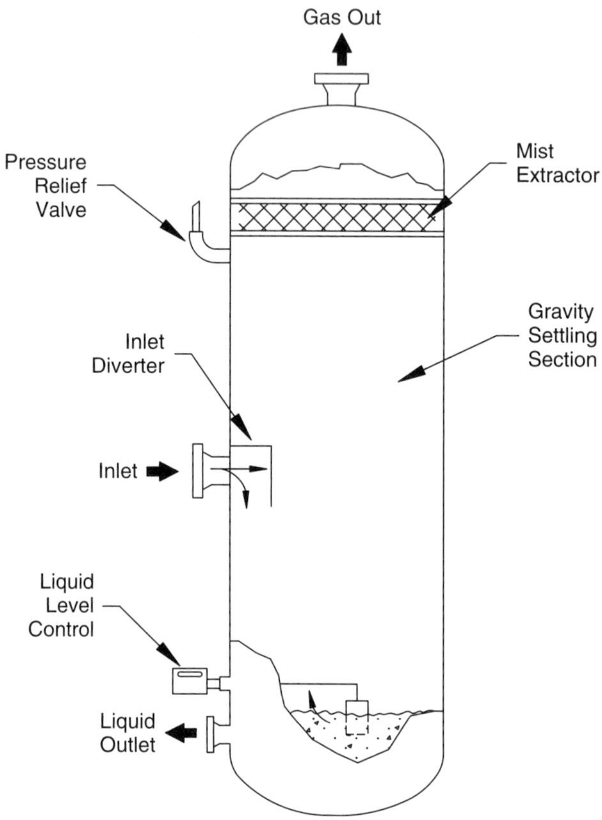

FIGURE 3.7. Cutaway view of a vertical two-phase separator.

Vertical separators are

- commonly used in flow streams with low to intermediate gas–liquid ratios,
- well suited for production containing sand and other sediments, and
- fitted with false cone bottom to handle sand production.

3.3.3 Spherical Separators

Figure 3.8 shows a typical spherical separator. The same four sections are found in this vessel. They are a special case of the vertical separator where there is not cylindrical shell between the two heads.

Fluid enters the vessel through the inlet diverter where the flow stream is split into two streams. Liquid falls to the liquid collection

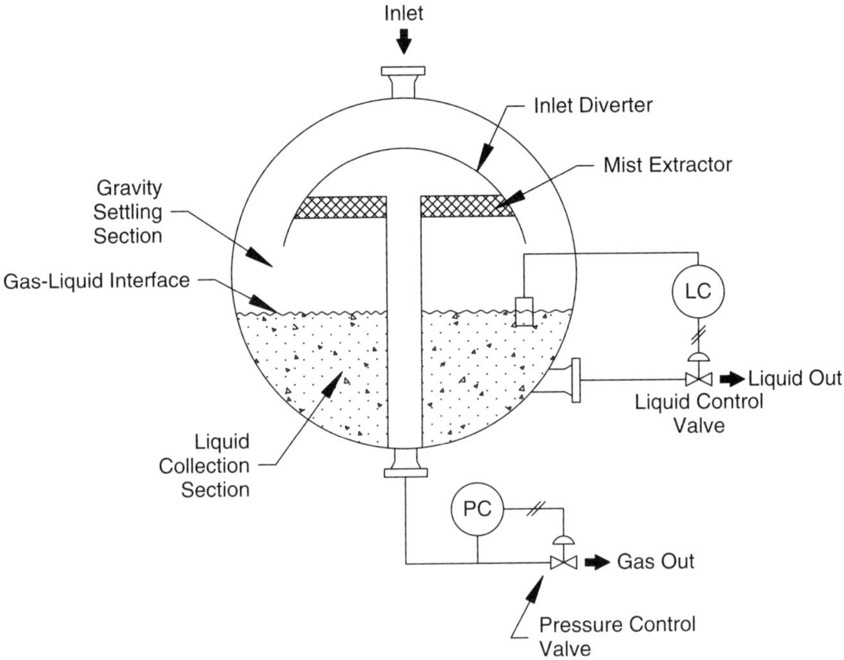

FIGURE 3.8. Spherical separator schematic.

section, through openings in a horizontal plate located slightly below the gas–liquid interface. The thin liquid layer across the plate makes it easier for any entrained gases to separate and rise to the gravity settling section.

Gas rising out of the liquids passes through the mist extractor and out of the separator through the gas outlet. Liquid level is maintained by a float connected to a dump valve. Pressure is maintained by a back pressure control valve, while liquid level is maintained by a liquid level dump valve.

Spherical separators were originally designed to take advantage, theoretically, of the best characteristics of both horizontal and vertical separators. In practice, however, these separators actually experienced the worst characteristics and are very difficult to size and operate. They may be very efficient from a pressure containment standpoint, but they are seldom used in oilfield facilities because

- They have limited liquid surge capability and
- they exhibit fabrication difficulties.

3.3.4 Centrifugal Separators

Centrifugal separators, sometimes referred to as a cylindrical cyclone separators (CCS), work on the principle that droplet separation can be enhanced by the importance of a radial or centrifugal force. Centrifugal force may range from 5 times the gravitational force in large-diameter units to 2500 times the gravitational force in small, high-pressure units.

As shown in Figure 3.9, the centrifugal separator consists of three major sections:

- inclined tangential inlet,
- tangential liquid outlet, and
- axial gas outlet.

The basis flow pattern involves a double vortex, with the gas spiraling downward along the wall and then upward in the center. The spiral velocity in the separator may reach several times the inlet velocity.

The flow patterns are such that the radial velocities are directed toward the walls, thus causing droplets to impinge on the vessel walls and run down to the bottom of the unit.

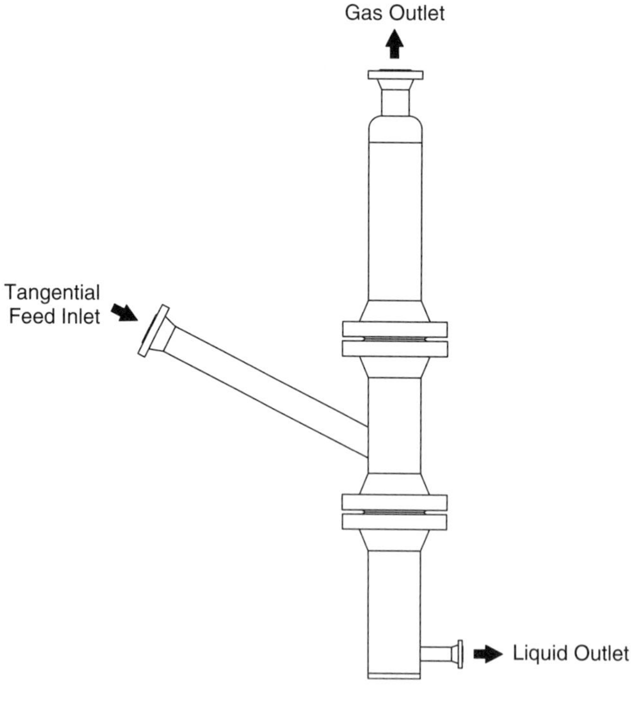

FIGURE 3.9. Cylindrical cyclone separator.

The units are designed to handle liquid flow rates between 100 and 50,000 bpd in sizes ranging from 2 to 12 in. diameter. Centrifugal separators are designed to provide bulk gas–liquid separations. They are best suited for fairly clean gas streams. Some of the major benefits are

- no moving parts,
- low maintenance,
- compact, in terms of weight and space,
- insensitive to motion, and
- low cost when compared to conventional separator technology.

They are not commonly used in production operations because

- their design is rather sensitive to flow rate, and
- they require greater pressure drop than the standard configurations previously described.

Since separation efficiency decreases as velocity decreases, the centrifugal separator is not suitable for widely varying flow rates. Units are commonly used to recover glycol carryover downstream of a glycol contact tower. The design of these separators is propriety and, therefore, will not be covered.

3.3.5 Venturi Separators

Like the centrifugal, the venturi separator increases droplet coalescence by introducing additional forces into the system. Instead of centrifugal forces, the venture acts on the principle of accelerating the gas linearly through a restricted flow path with a motive fluid to promote the coalescence of droplets.

Venturi separators are

- best suited for application that contain a mixture of solids and liquids and
- not cost effective for removing liquid entrainment alone, because of the high-pressure drop and need for a motive fluid.

Even with solids present, the baffle-type units are more suitable for entrained particulars down to 15 μm in diameter.

3.3.6 Double-Barrel Horizontal Separators

Figure 3.10 illustrates a double-barrel horizontal separator, which is a variation of the horizontal separator. The gas and liquid chambers are separated.

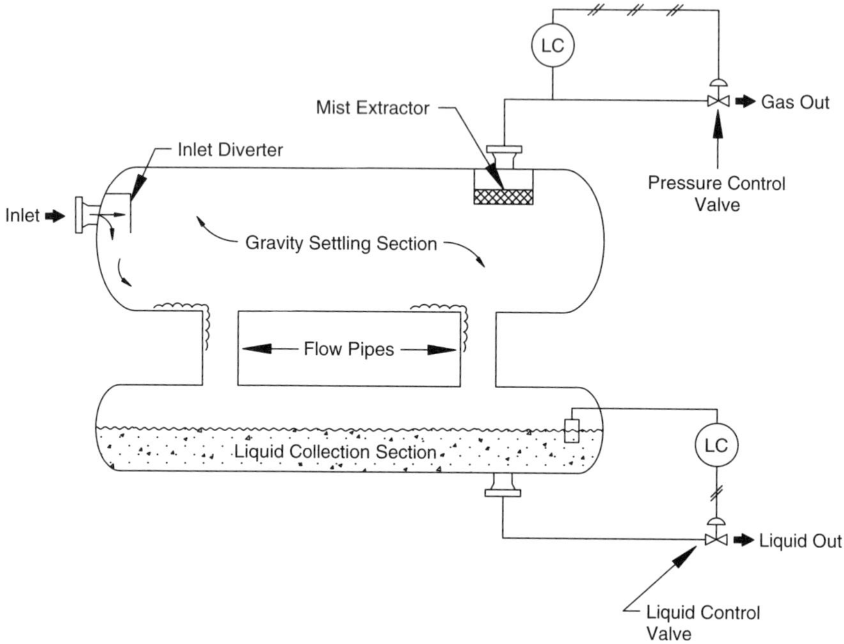

FIGURE 3.10. Double-barrel horizontal separator.

These are commonly used in applications where there are high gas flow rates and where there is a possibility of large slugs—for example, slug catchers. Single-barrel horizontal separators can handle large gas flow rates but offer poor liquid surge capabilities.

Flow stream enters the vessel in the upper barrel and strikes the inlet diverter. The gas flows through the gravity settling section, where it encounters the baffling-type mist extractors enroute to the gas outlet.

Figure 3.11 is a cutaway view of a double-barrel separator fitted with a baffle-type mist extractor. Baffles help the free liquids to fall to the lower barrel through flow pipes. Liquids drain through the flow pipe into the lower barrel.

Small amounts of gas entrained in the liquid are liberated in the liquid collection barrel and flow up through the flow pipes. These are not widely used in oil field systems because of

- additional cost and
- absence of problems with single-vessel separators.

These are typically used in gas handling, conditioning, and processing facilities as gas scrubbers on the inlet of compressors, glycol contact

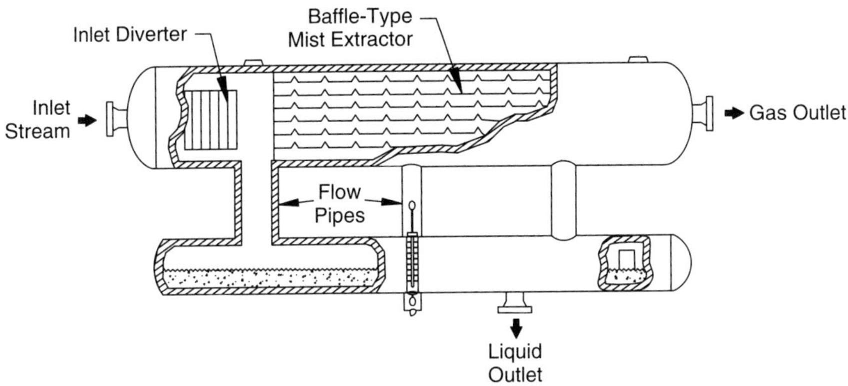

FIGURE 3.11. Cutaway view of a horizontal double-barrel separator fitted with a baffle-type mist extractor in the gravity settling section.

towers, and gas treating systems where the liquid flow rate is extremely low relative to the gas flow rate.

3.3.7 Horizontal Separator with a Boot or Water Pot

Figure 3.12 shows a special case of a two-barrel separator. It is a single-barrel separator with a liquid "boot" or "water pot" at the outlet

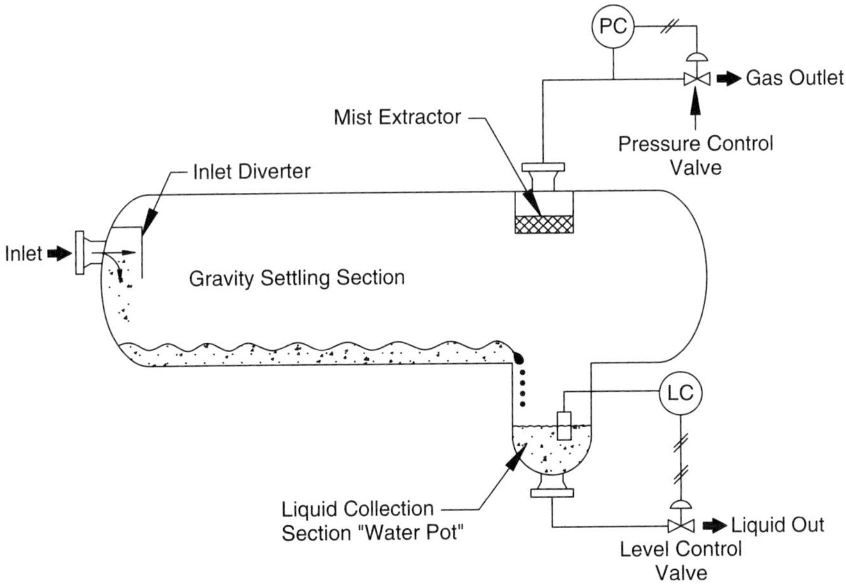

FIGURE 3.12. Single-barrel horizontal separator with a liquid boot.

end. The main body of the separator operates essentially dry as in a two-barrel separator. The small amounts of liquid in the bottom flow to the boot end, which provides the liquid collection section. These vessels are less expensive than two-barrel separators, but they also contain less liquid-handling capacity. It is used where there are very low liquid flow rates, especially where the flow rates are low enough that the boot can serve as a liquid–liquid separator as well.

3.3.8 Filter Separator

The filter separator is frequently used in some high-gas/low-liquid flow applications. It is designed to remove small liquid and solid particles from the gas stream. These are used in applications where conventional separators employing gravitational or centrifugal force are ineffective.

Figure 3.13 shows a horizontal two-barrel filter separator design. Filter tubes in the initial separation section cause coalescence of any liquid mist into larger droplets as the gas passes through the tubes.

A secondary section of vanes or other mist extractor elements removes these coalesced droplets. They are commonly used on compressor inlets in field compressor stations, final scrubbers upstream of glycol contact towers, and instrument/fuel gas applications. Design is propriety and dependent on the type of filter element employed.

Some elements can remove 100% of 1-μm particles and 99% of ½-μm particles when they are operated at rated capacity and recommended filter-change intervals.

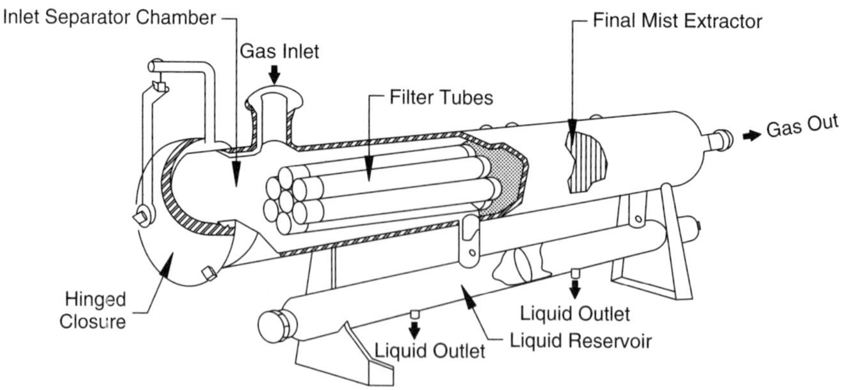

FIGURE 3.13. Typical horizontal two-barrel filter separator.

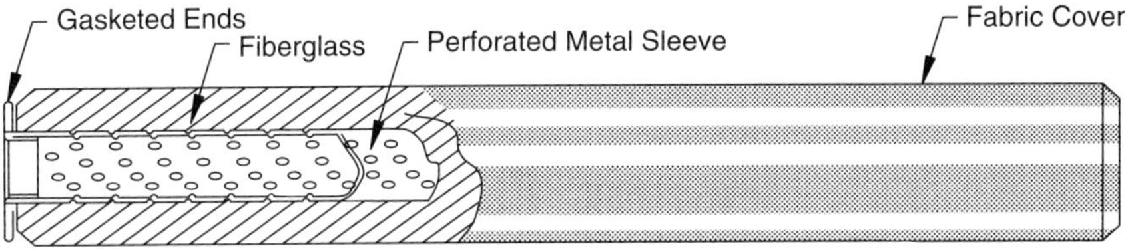

FIGURE 3.14. Typical filter element.

Figure 3.14 shows a typical filter element. The element consists of

- a perforated metal cylinder with gasketed ends for compression sealing and
- a fiberglass cylinder, typically ½-in. (1.25-cm) thick, surrounds the perforated metal cylinder.

3.3.9 Scrubbers

A scrubber is a two-phase separator that is designed to recover liquids carried over from the gas outlets of production separators or to catch liquids condensed due to cooling or pressure drops. Liquid loading is much lower than that in a separator. Typical applications include:

- upstream of mechanical equipment such as compressors that could be damaged, destroyed, or rendered ineffective by free liquid;
- downstream of equipment that can cause liquids to condense from a gas stream (such as coolers);
- upstream of gas dehydration equipment that would lose efficiency, be damaged, or destroyed if contaminated with liquid hydrocarbons; and
- upstream of a vent or flare outlet.

Vertical scrubbers are most commonly used. Horizontal scrubbers can be used, but space limitations usually dictate the use of a vertical configuration.

3.3.10 Slug Catchers

A "slug catcher," commonly used in gas gathering pipelines, is a special case of a two-phase gas–liquid separator that is designed to handle large gas capacities and liquid slugs on a regular basis. Figure 3.15 is a schematic of a two-phase horizontal slug catcher with liquid "fingers."

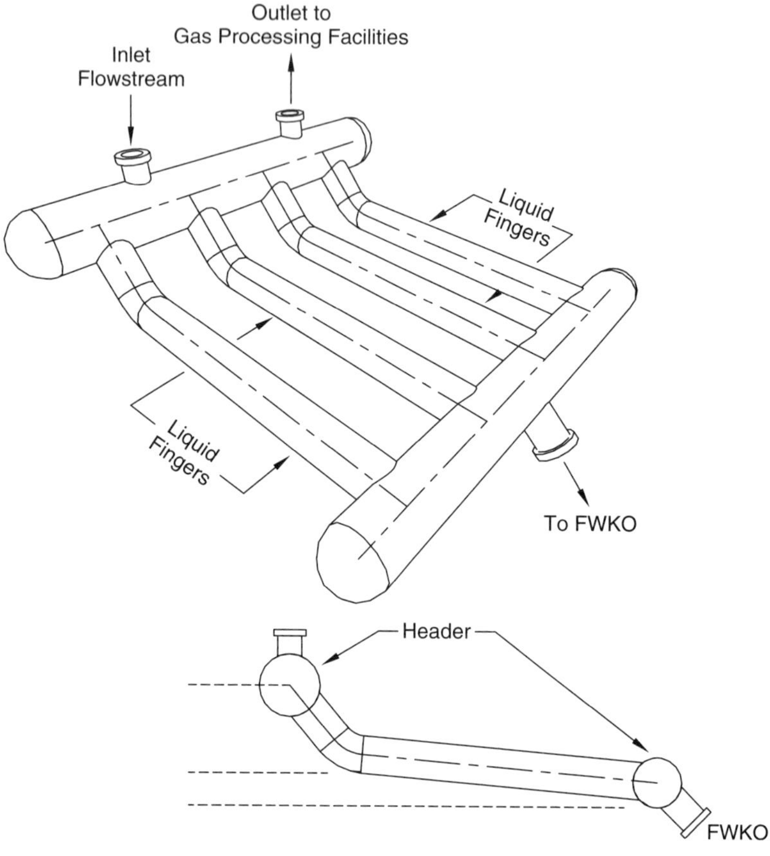

FIGURE 3.15. Schematic of a two-phase horizontal slug catcher with liquid fingers.

Gas and liquid slug from the gathering system enters the horizontal portion of the two-phase vessel, where primary gas–liquid separation is accomplished. Gas exits the top of the separator through the mist extractor, while the liquid exits the bottom of the vessel through a series of large-diameter tubes, or fingers. The tubes provide a large liquid holding volume and route the liquid to a three-phase free water knockout (FWKO) for further liquid–liquid separation.

3.4 Selection Considerations

The geometry of and physical and operating characteristics give each separator type advantages and disadvantages.

Horizontal separators are

- smaller,
- more efficient at handling large volumes of gas, and
- less expensive than vertical separators for a given gas capacity.

In the gravity settling section of a horizontal vessel, the liquid droplets fall perpendicularly to the gas flow and thus are more easily settled out of the gas continuous phase.

Since the interface area is larger in a horizontal separator than a vertical separator, it is easier for the gas bubbles, which come out of solution as the liquid approaches equilibrium, to reach the vapor space.

Horizontal separators offer greater liquid capacity and are best suited for liquid–liquid separation and foaming crude. Horizontal separators

- are not as good as vertical separators in handling solids,
- require more plan area to perform the same separation as vertical vessels, and
- can have less liquid surge capacity than vertical vessels sized for the same steady-state flow rate.

Since vertical separators are supported only by the bottom skirt (Figure 3.16), the walls of vertical separators must be somewhat thicker than a similarly sized and rated horizontal separator, which may be supported by saddles.

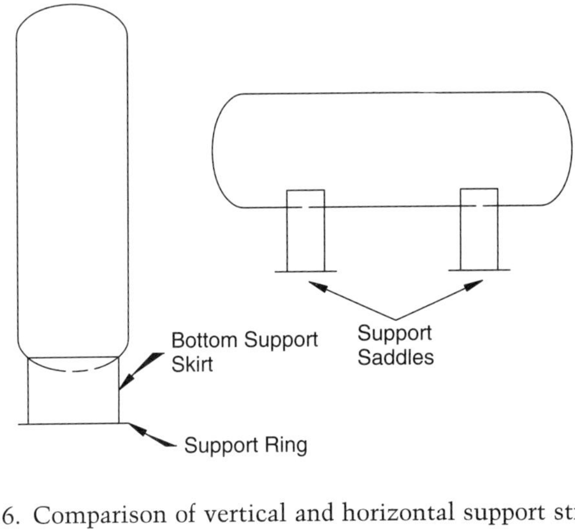

FIGURE 3.16. Comparison of vertical and horizontal support structures.

Overall, horizontal vessels are the most economical for normal oil–gas separation, particularly where there may be problems with emulsions, foam, or high gas–oil ratios (GOR).

Vertical vessels work most effectively in low-GOR applications. They are also used in some very high GOR applications, such as scrubbers where only fluid mists are being removed from the gas and where extra surge capacity is needed to allow shutdown to activate before the liquid is carried out of the gas outlet (e.g., compressor suction scrubber).

3.5 Vessel Internals

3.5.1 Inlet Diverters

Inlet diverters serve to impart flow direction of the entering vapor/liquid stream and provide primary separator between the liquid and vapor. There are many types of inlet diverters. Three main types are baffle plates (shown in Figure 3.17), centrifugal diverters (shown in Figure 3.18), and elbows (shown in Figure 3.19).

A baffle plate can be a spherical dish, flat plate, angle iron, cone, elbow, or just about anything that will accomplish a rapid change in direction and velocity of the fluids and thus disengage the gas and liquid. At the same velocity the higher-density liquid possesses more energy and thus does not change direction or velocity as easily as the gas. Thus, the gas tends to flow around the diverter while the liquid strikes the diverter and then falls to the bottom of the vessel.

The design of the baffles is governed principally by the structural supports required to resist the impact-momentum load. The advantage of using devices such as a half-sphere elbow or cone is that they

FIGURE 3.17. Baffle plates.

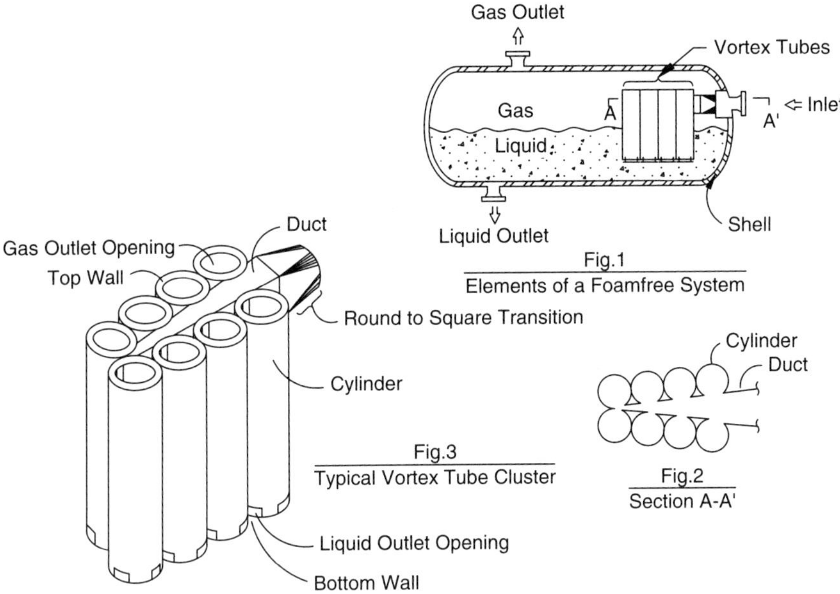

FIGURE 3.18. Three views of an example centrifugal inlet diverter (courtesy of Porta-Test Systems, Inc.).

create less disturbance than plates or angle iron, cutting down on re-entrainment or emulsifying problems.

Centrifugal inlet diverters use centrifugal force, rather than mechanical agitation, to disengage the oil and gas. These devices can have a cyclonic chimney or may use a tangential fluid race around the walls (Figure 3.20).

Centrifugal inlet diverters are proprietary but generally use an inlet nozzle sufficient to create a fluid velocity of about 20 f/s (6 m/s) around a chimney whose diameter is no longer than two-thirds that of the vessel diameter. Centrifugal diverters can be designed to efficiently separate the liquid while minimizing the possibility of foaming or emulsifying problems.

The disadvantage is that their design is rate sensitive. At low velocities they will not work properly. Thus, they are not normally recommended for producing operations where rates are not expected to be steady.

3.5.2 Wave Breakers

In long, horizontal vessels, usually located on floating structures, it may be necessary to install wave breakers. The waves may result from surges

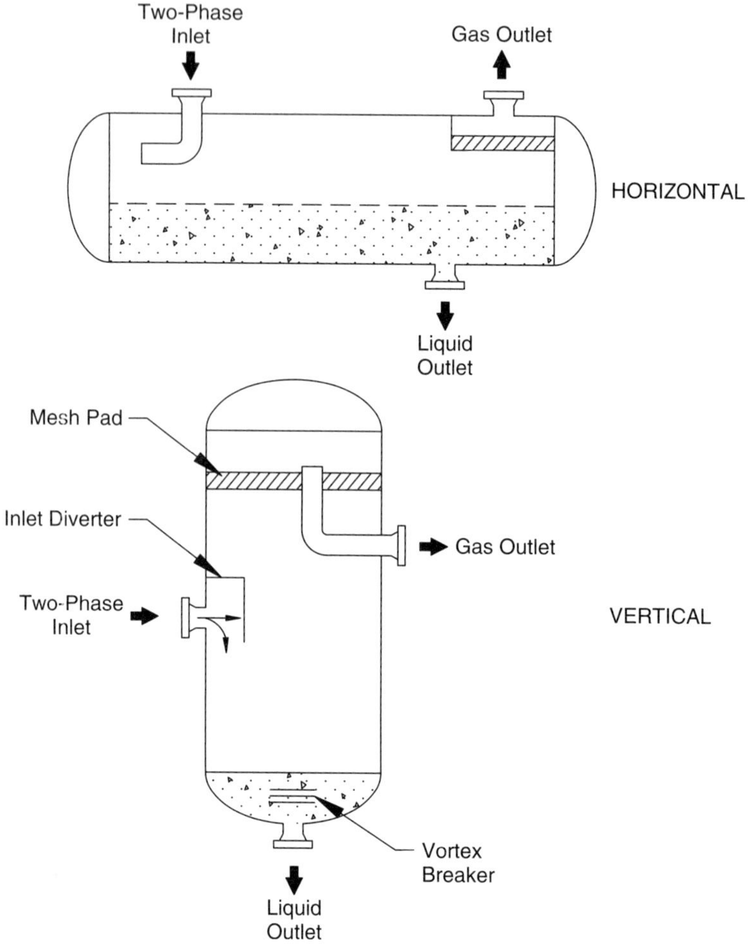

FIGURE 3.19. Elbow inlet diverter.

of liquids entering the vessel. Wave breakers are nothing more than perforated baffles or plates that are placed perpendicularly to the flow located in the liquid collection section of the separator. These baffles dampen any wave action that may be caused by incoming fluids.

On floating or compliant structures where internal waves may be set up by the motion of the foundation, wave breakers may also be required perpendicular to the flow direction. The wave actions in the vessel must be eliminated so level controls, level switches, and weirs may perform properly. Figure 3.21 is a three-dimensional view of a horizontal separator fitted with an inlet diverter, defoaming element, mist extractor, and wave breakers.

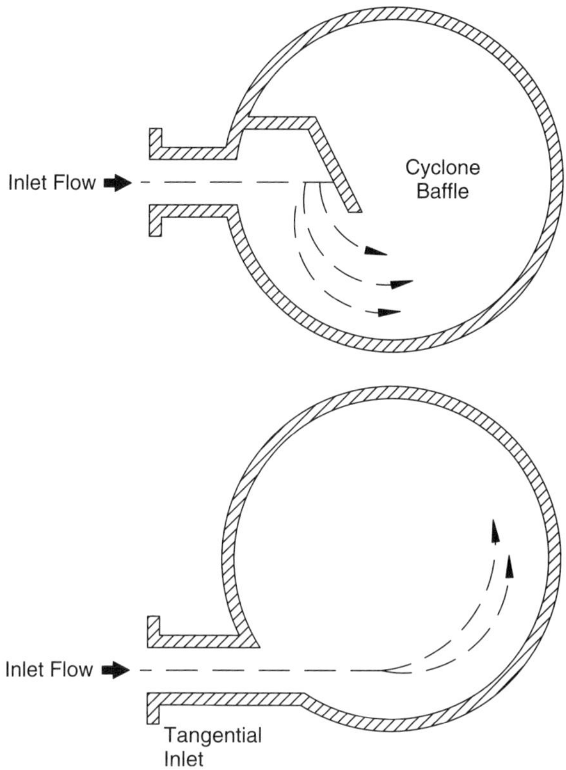

FIGURE 3.20. Centrifugal inlet diverters. (Top) Cyclone baffle. (Bottom) Tangential raceway.

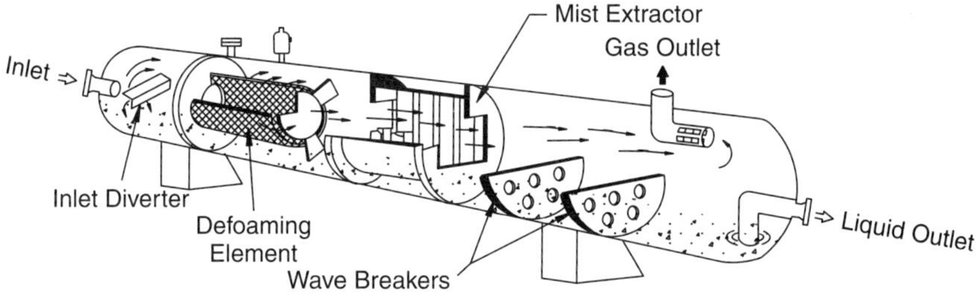

FIGURE 3.21. Three-dimensional view of a horizontal separator fitted with an inlet diverter, defoaming element, mist extractor, and wave breaker.

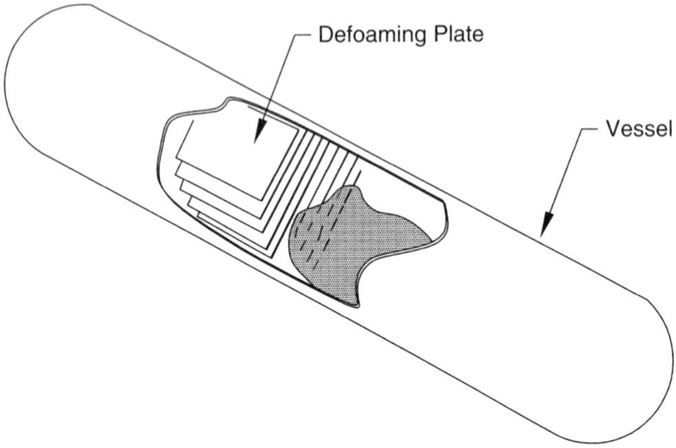

FIGURE 3.22. Defoaming plates.

3.5.3 Defoaming Plates

Foam at the interface may occur when gas bubbles are liberated from the liquid. Foam can severely degrade the performance of a separator. This foam can be stabilized with the addition of chemicals at the inlet. Many times a more effective solution is to force the foam to pass through a series of inclined parallel plates or tubes as shown in Figure 3.22. These closely spaced, parallel plates or tubes provide additional surface area, which breaks up the foam and allows the foam to collapse into the liquid layer.

3.5.4 Vortex Breaker

Liquid leaving a separator may form vortices or whirlpools, which can pull gas down into the liquid outlet. Therefore, horizontal separators are often equipped with vortex breakers, which prevent a vortex from developing when the liquid control valve is open. A vortex could suck some gas out of the vapor space and re-entrain it in the liquid outlet.

One type of vortex breaker is shown in Figure 3.23. It is a covered cylinder with radially directed flat plates. As liquid enters the bottom of the vortex breaker, any circular motion is prevented by the flat plates. Any tendency to form vortices is removed. Figure 3.24 illustrates other commonly used vortex breakers.

3.5.5 Stilling Well

A stilling well, which is simply a slotted pipe fitting surrounding an internal level control displacer, protects the displacer from currents, waves, and other disturbances that could cause the displacer to sense an incorrect level measurement.

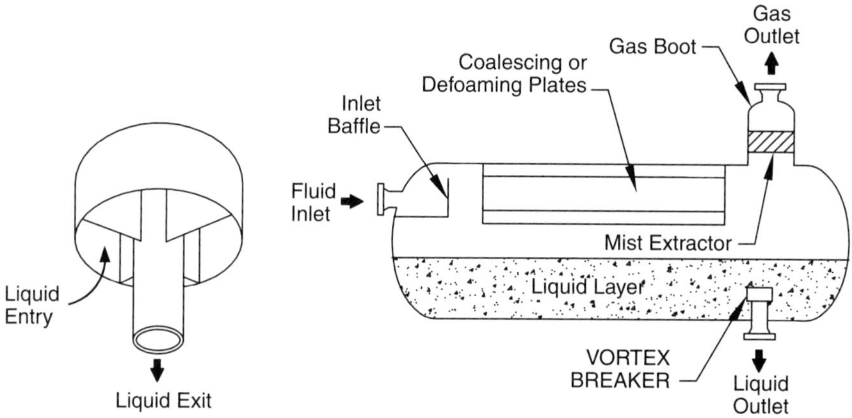

FIGURE 3.23. Vortex breaker.

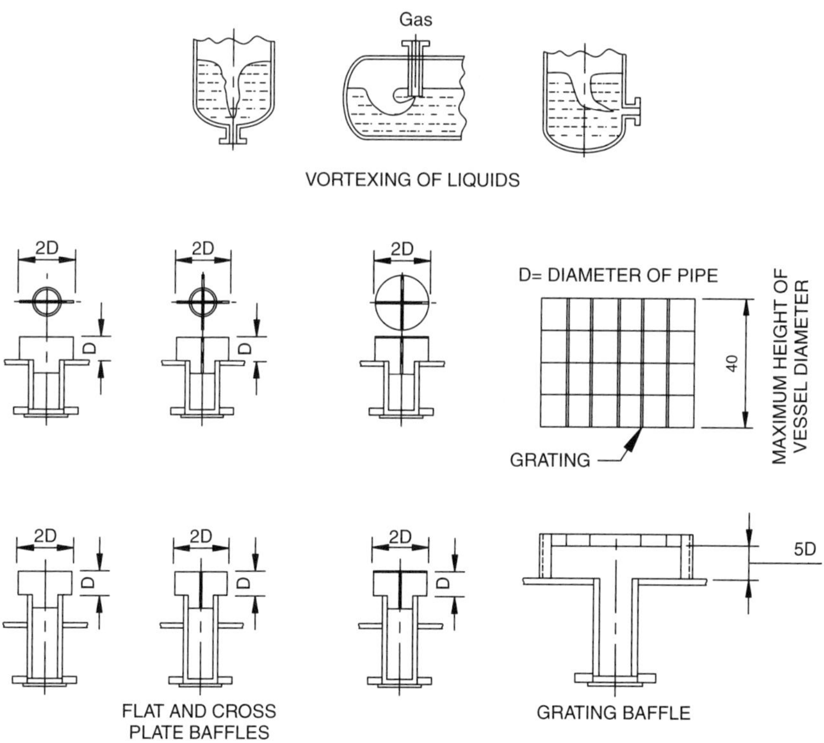

FIGURE 3.24. Typical vortex breakers.

3.5.6 Sand Jets and Drains

In horizontal separators, one worry is the accumulation of sand and solids at the bottom of the vessel. If allowed to build up, these solids will upset the separator operations by taking up vessel volume. Generally, the solids settle to the bottom and become well packed.

To remove the solids, sand drains are opened in a controlled manner, and then high-pressure fluid, usually produced water, is pumped through the jets to agitate the solids and flush them down the drains. The sand jets are normally designed with a 20-ft/s (6-m/s) jet tip velocity and aimed in such a manner to give good coverage of the vessel bottom.

To prevent the settled sand from clogging the sand drains, sand pans or sand troughs are used to cover the outlets. These are inverted troughs with slotted side openings (Figure 3.25).

To ensure proper solids removal without upsetting the separation process, an integrated system, consisting of a drain and its associated jets, should be installed at intervals not exceeding 5 ft (1.5 m). Field experience indicates it is not possible to mix and fluff the bottom of a long, horizontal vessel with a single sand jet header.

3.5.7 Mist Extractors

Introduction

There are many types of equipment known as mist extractors or mist eliminators, which are designed to remove the liquid droplets and

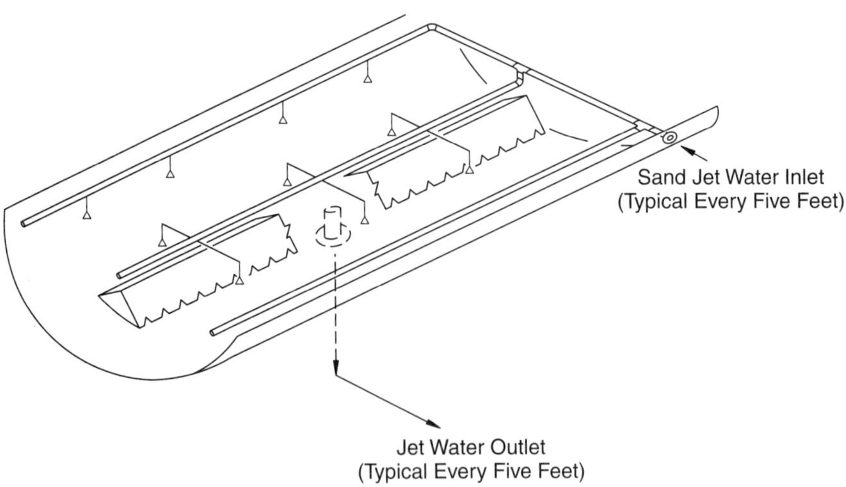

FIGURE 3.25. Schematic of a horizontal separator fitted with sand jets and inverted trough.

solid particles from the gas stream. Before a selection can be made, one must evaluate the following factors:

- Size of droplets the separator must remove.
- Pressure drop that can be tolerated in achieving the required level of removal.
- Susceptibility of the separator to plugging by solids, if solids are present.
- Liquid handling capability of the separator.
- Whether the mist extractor/eliminator can be installed inside existing equipment, or if it requires a standalone vessel instead.
- Availability of the materials of construction that are comparable with the process.
- Cost of the mist extractor/eliminator itself and required vessels, piping, instrumentation, and utilities.

Gravitational and Drag Forces Acting on a Droplet

All mist extractor types are based on the same kind of intervention in the natural balance between gravitational and drag forces. This is accomplished in one or more of the following ways:

- Overcoming drag force by reducing the gas velocity (gravity separators or settling chambers)
- Introducing additional forces (venturi scrubbers, cyclones, electrostatic precipitators)
- Increasing gravitational force by boosting the droplet size (impingement-type)

The relevant laws of fluid mechanics and the principal forces acting on a liquid droplet falling through the continuous gas phase are discussed below. As the gas in a vessel flows upward, there are two opposing forces acting on a liquid droplet: a gravitational force (or negative buoyant force) acting downward to accelerate the droplet and an opposing drag force acting to slow the droplet's rate of fall. An increase in the upward gas velocity increases the drag force on the droplet. The drag force continues to reduce the rate of fall until a point is reached when the downward velocity reaches zero, and the droplet becomes stationary. When the gravitational or negative buoyant force equals the drag force, the acceleration of the liquid droplet becomes zero and the droplet will settle at a constant "terminal" or "settling" velocity. Additional increases in gas velocity result in an initial reduction in settling velocity of the droplet. Further increase causes the droplet to move upward at increasing velocities until a point is reached where the droplet velocity approaches the gas velocity. The same theory is applicable to horizontal gas flow as well. The primary difference is that the gravitational and drag forces are operating at 90° to each other. Thus, there is always a net force acting in the downward direction.

Impingement-type

The most widely used type of mist extractor is the impingement-type because it offers good balance among efficiency, operating range, pressure drop requirement, and installed cost. These types consist of baffles, wire meshes, and microfiber pads.

Impingement-type mist extractors may involve just a single baffle or disc installed in a vessel. As illustrated in Figure 3.26, as the gas approaches the surface of the baffle or disc (commonly referred to as a target), fluid streamlines spread around the baffle or disc. Ignoring the eddy streams formed around the target, one can assume that the higher the stream velocity, the closer to the target these streamlines start to form. A droplet can be captured by the target in an impingement-type mist extractor/eliminator via any of the following three mechanisms: inertial impaction, direct interception, and diffusion (Figure 3.26).

- *Inertial impaction:* Because of their mass, particles 1–10 μm in diameter in the gas stream have sufficient momentum to break

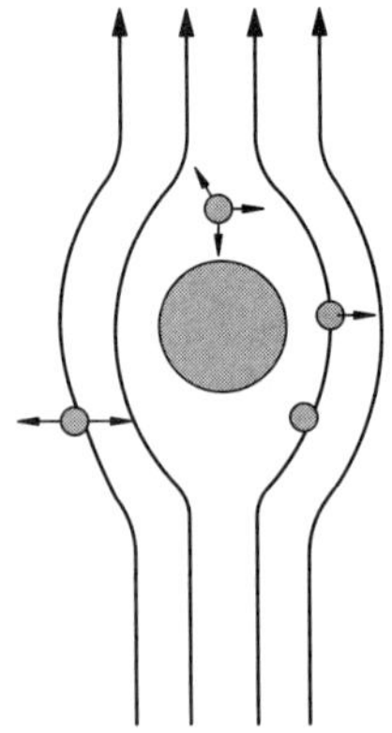

FIGURE 3.26. The three primary mechanisms of mist capture via impingement are inertial impaction (left), direct interception (center), and Brownian diffusion (right).

through the gas streamlines and continue to move in a straight line until they impinge on the target. Impaction is generally the most important mechanism in wire-mesh pads and impingement plates.

- *Direct interception:* There are also particles in the gas stream that are smaller, between 0.3 and 1 μm in diameter, than those above. These do not have sufficient momentum to break through the gas streamlines. Instead, they are carried around the target by the gas stream. However, if the streamline in which the particle is traveling happens to lie close enough to the target so that the distance from the particle centerline to the target is less than one-half the particle's diameter, the particle can touch the target and be collected. Interception effectiveness is a function of pore structure. The smaller the pores, the greater the media to intercept particles.
- *Diffusion:* Even smaller particles, usually smaller than 0.3 μm in diameter, exhibit random Brownian motion caused by collisions with the gas molecules. This random motion will cause these small particles to strike the target and be collected, even if the gas velocity is zero. Particles diffuse from the streamlines to the surface of the target where the concentration is zero. Diffusion is favored by low-velocity and high-concentration gradients.

Baffles

This type of impingement mist extractor consists of a series of baffles, vanes, or plates between which the gas must flow. The most common is the vane or chevron-shape, as shown in Figures 3.27 and 3.28.

The vanes force the gas flow to be laminar between parallel plates that contain directional changes. The surface of the plates serves as a target for droplet impingement and collection. The space between the baffles ranges from 5 to 75 mm, with a total depth in the flow direction of 150–300 mm.

Figures 3.29 and 3.30 illustrate a vane mist extractor installed in a vertical and horizontal separator, respectively. Figure 3.31 shows a vane mist extractor made from an angle iron. Figure 3.32 illustrates an "arch" plate mist extractor. As gas flows through the plates, droplets impinge on the plate surface. The droplets coalesce, fall, and are routed to the liquid collection section of the vessel. Vane-type eliminators are sized by their manufacturers to ensure both laminar flow and a certain minimum pressure drop. Vane or chevron-shaped mist extractors remove liquid droplets 10–40 μm and

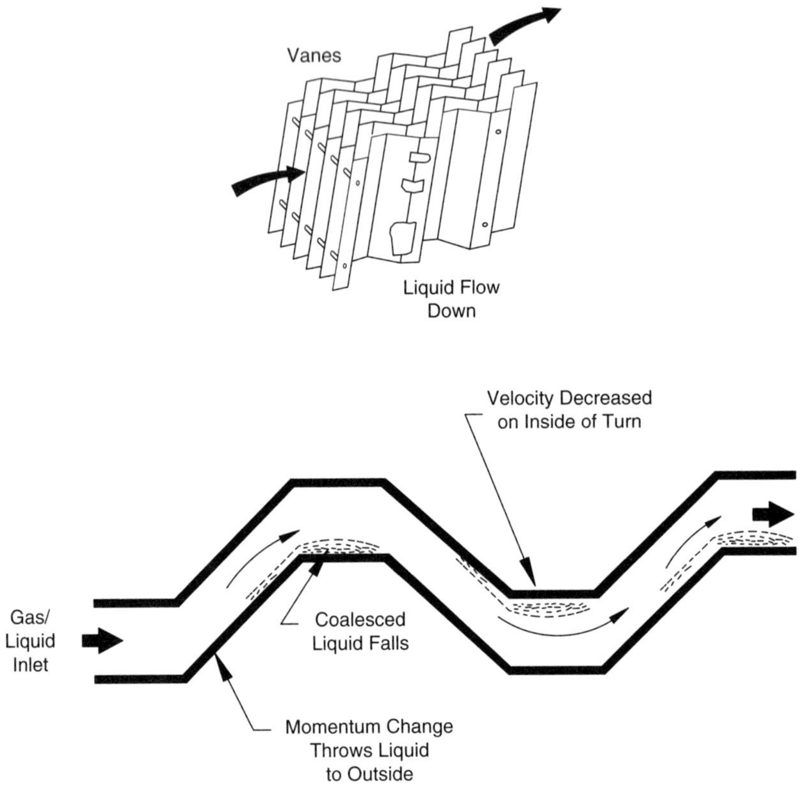

FIGURE 3.27. Typical vane-type mist extractor/eliminator.

larger. Their operation is usually dictated by a design velocity expressed as follows:

$$V = K\sqrt{\left[\frac{(\rho_l - \rho_g)}{\rho_l}\right]} \tag{3.1}$$

where V = gas velocity, K = Souders–Brown coefficient, ρ_l = liquid or droplet density, and ρ_g = gas density.

The K factor, or Souders–Brown coefficient, is determined experimentally for each plate geometry. Its value ranges from 0.3 to 1.0 ft/s (0.09–0.3 m/s) in typical designs.

Since impaction is the primary collection mechanism, at too low a value of K, the droplets can remain in the gas streamlines and pass through the device uncollected. The upper limit is set to minimize re-entrainment, which is caused either by excessive breakup of the droplets as they impinge onto the plates or by shearing of the liquid film on the plates.

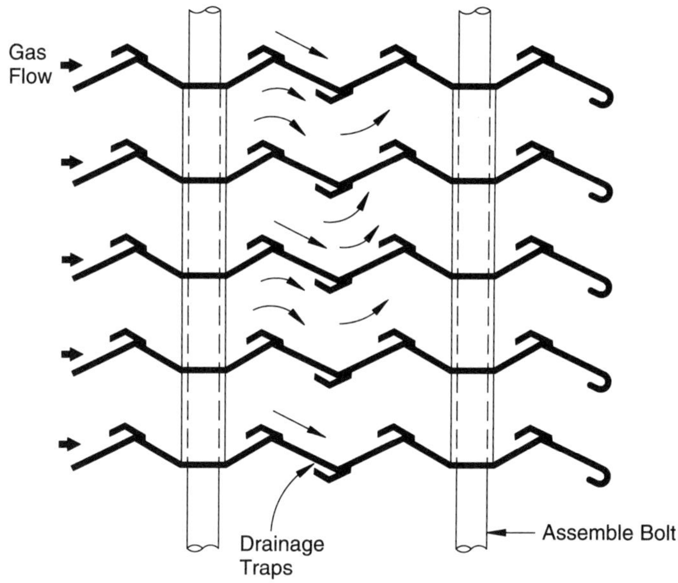

FIGURE 3.28. Vane-type element with corrugated plates and liquid drainage trays.

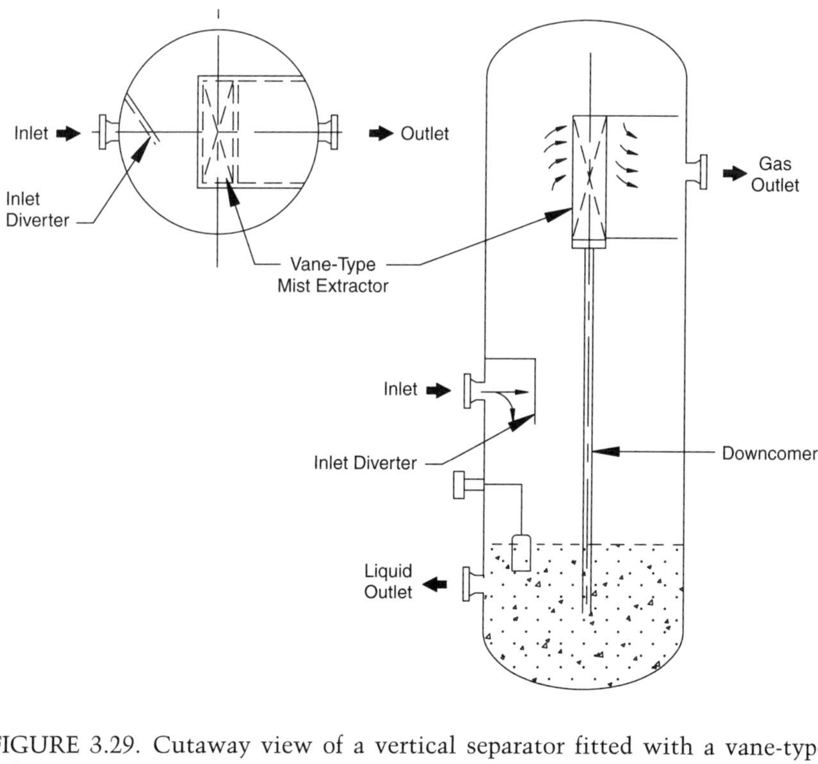

FIGURE 3.29. Cutaway view of a vertical separator fitted with a vane-type mist extractor.

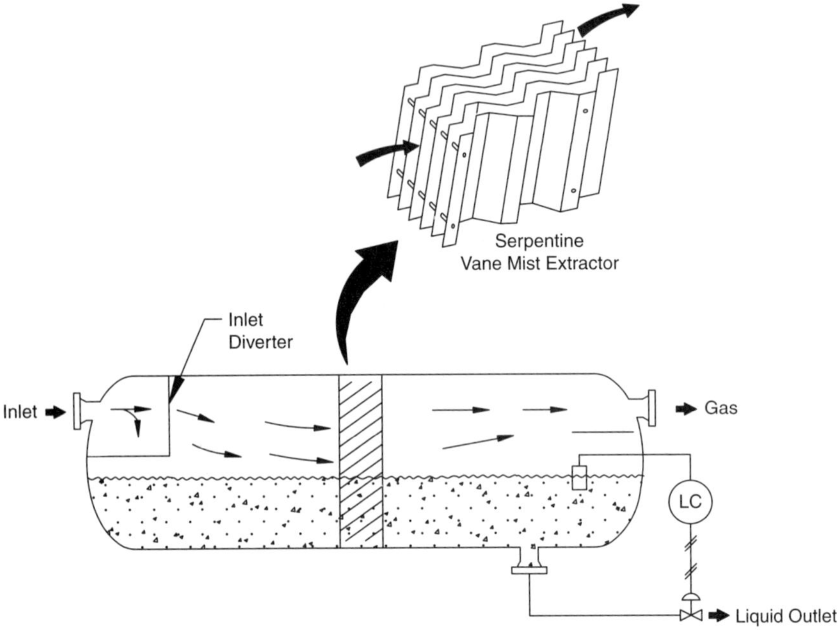

FIGURE 3.30. Cutaway view of a horizontal separator fitted with a vane-type mist extractor.

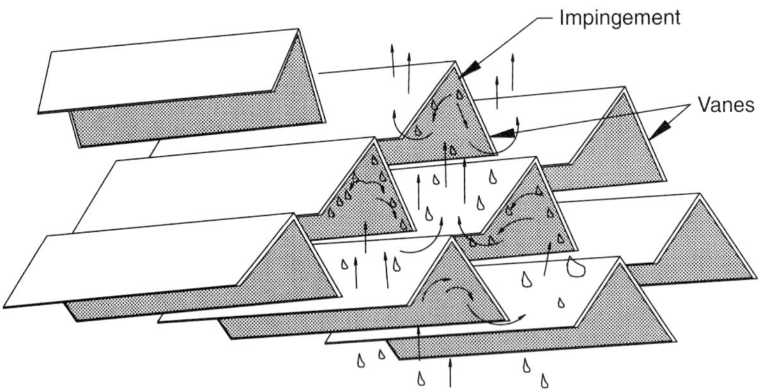

FIGURE 3.31. A vane-type mist extractor made from angle iron.

Higher gas velocities can be handled if the vanes are installed in a horizontal gas flow instead of vertical up-flow. In the horizontal configuration the liquid can easily drain downward due to gravity and thus out of the path of the incoming gas, which minimizes re-entrainment of the liquid.

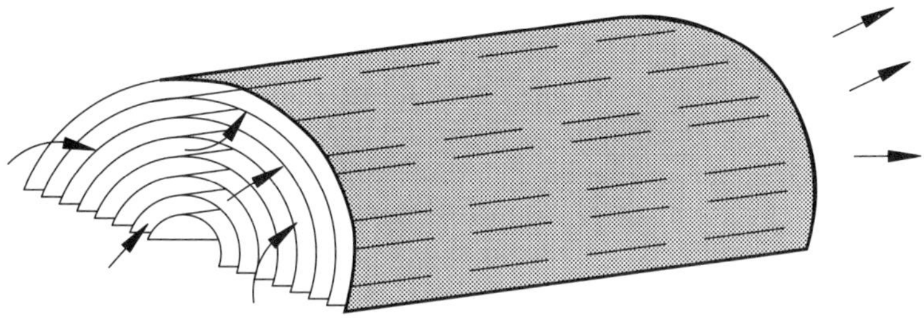

FIGURE 3.32. An arch plate-type mist extractor.

The vane type appears most often in process systems, where the liquid entrainment is contaminated with solids or where high liquid loading exists. Vane-type mist extractors are less efficient in removing very small droplets than other impaction types such as wire-mesh or microfiber. Standard designs are generally limited to droplets larger than 40 μm. However, high-efficiency designs provide droplet removal down to less than 15 μm in diameter. The pressure drop is low, often less than 10–15 mmH_2O.

Wire-mesh

The most common type of mist extractor found in production operations is the knitted-wire-mesh type (Figure 3.33). These units outnumber all other types of mist extractors. They are knitted (rather than woven) wire, and these devices have high surface area and void volume. Whereas woven wire has one set of wires running perpendicularly to a second set of wires, knitted wire instead has a series of interlocking loops just like cloth fiber. This makes the knitted product sufficiently flexible and yet structurally stable.

FIGURE 3.33. Example wire-mesh mist extractor (photo courtesy of ACS Industries, LP, Houston, TX).

The wire-mesh mist extractor is often specified by calling for a certain thickness (usually 3–7 in.) and mesh density (usually 10–12 lb/ft^3). They are usually constructed from wires of diameter ranging from 0.10 to 0.28 mm, with a typical void volume fraction of 0.95–0.99. The wire pad is placed between top and bottom support grids to complete the assembly. The grids must be strong enough to span between the supports and have sufficient free area for flow. Wire-mesh pads are mounted near the outlet of a separator, generally on a support ring (vertical separator) or frame (horizontal separator; cf. Figures 3.34 and 3.35, respectively).

Wire-mesh mist extractors are normally installed in vertical upward gas flow, although horizontal flows are employed in some specialized applications. In a horizontal flow the designer must be careful because liquid droplets captured in the higher elevation of the vertical mesh may drain downward at an angle as they are pushed through the mesh, resulting in re-entrainment.

The effectiveness of wire-mesh depends largely on the gas being in the proper velocity range [Equation (3.1)]. If the velocities are too high, the liquids knocked out will be re-entrained. If the velocities are low, the vapor just drifts through the mesh element without the droplets impinging and coalescing. The lower limit of the velocity is normally set at 30% of design velocity, which maintains a reasonable efficiency. The upper limit is governed by the need to prevent re-entrainment of liquid droplets from the downstream face of the wire-mesh device.

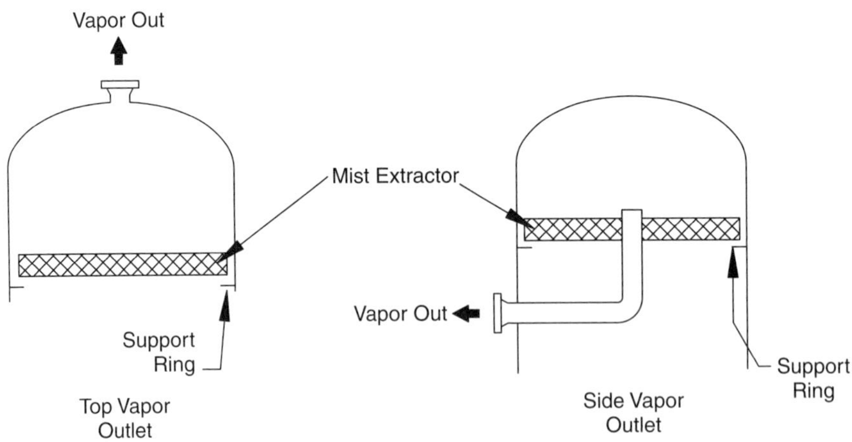

FIGURE 3.34. Vertical separators fitted with wire-mesh pads supported by support rings.

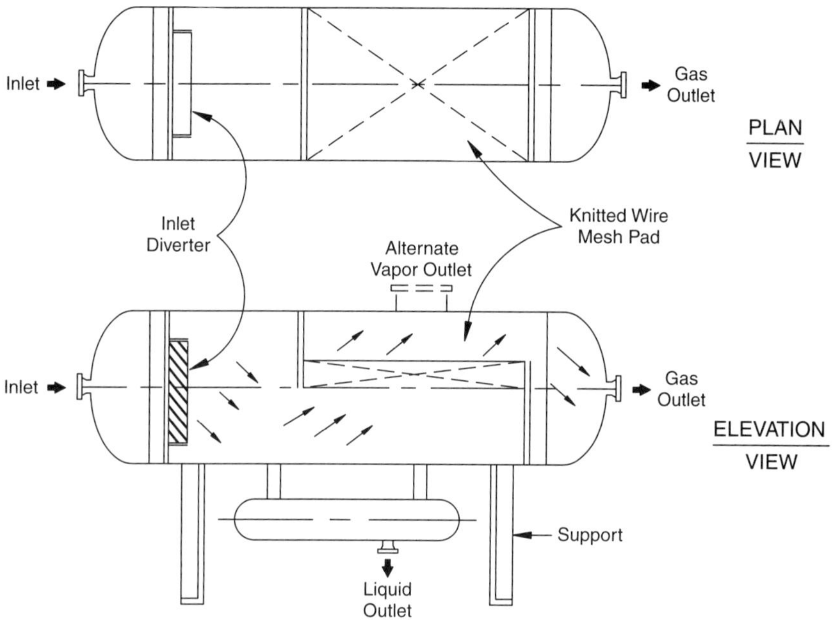

FIGURE 3.35. Horizontal separator fitted with wire-mesh pads supported by a frame.

The pressure drop through a wire-mesh unit is a combination of "dry" pressure drop due to gas flow only, plus the "wet" pressure drop due to liquid holdup. The dry pressure drop may be calculated from the following equation:

$$\Delta P_{\text{dry}} = \frac{fHa\rho_g V^2}{981 \times 10^{30}} \tag{3.2}$$

where f = friction factor from Figure 3.36; H = thickness of mesh pad, in.; a = surface area, in.^{2}; ρ_g = gas density, lb/ft^3; V = gas velocity, ft/s; and ΔP_{dry} = pressure drop, psi.

The wet pressure drop, a function of liquid loading as well as wire-mesh pad geometry, may be obtained experimentally over a range of gas velocities and liquid loadings. There are also correlations available for the various wire-mesh geometries.

Whether installed inside a piece of process equipment or placed inside a separate vessel of its own, a wire-mesh or baffle-type mist extractor offers low-pressure drop. To ensure a unit's operation at design capacity and high mist elimination efficiency, the flow pattern of the gas phase must be uniform throughout the element.

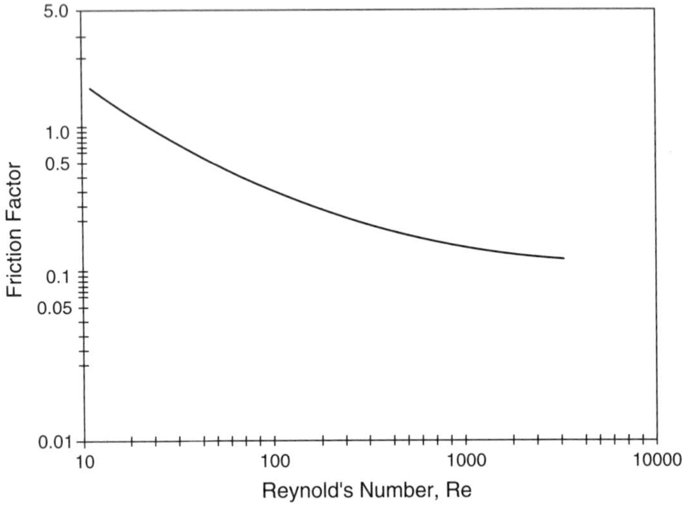

FIGURE 3.36. Friction factor versus Reynolds number for a dry knitted wire-mesh extractor.

When there are size limitations inside a process vessel, an integral baffle plate can be used on the downstream side face of the wire-mesh element as a vapor distributor. Even here the layout of the drum must be such that the flow stream enters the mesh pad with flow-pattern streamlines that are nearly uniform.

When knockout drums are equipped with vanes or wire-mesh pads, one can use any one of the four following design configurations: horizontal or vertical vessels, with horizontal or vertical vane or mesh elements. The classic configuration is the vertical vessel with horizontal element. In order to achieve uniform flow, one has to follow a few design criteria (Figure 3.37).

A properly sized wire-mesh unit can remove 100% of liquid droplets larger than 3–10 μm in diameter. Although wire-mesh eliminators are inexpensive, they are more easily plugged than the other types. Wire-mesh pads are not the best choice if solids can accumulate and plug the pad.

Microfiber

Microfiber mist extractors use very small diameter fibers, usually less than 0.02 mm, to capture very small droplets. Gas and liquid flow is horizontal and co-current. Because the microfiber unit is manufactured from densely packed fiber, drainage by gravity inside the unit is limited. Much of the liquid is eventually pushed through the microfiber and drains on the downstream face. The surface area

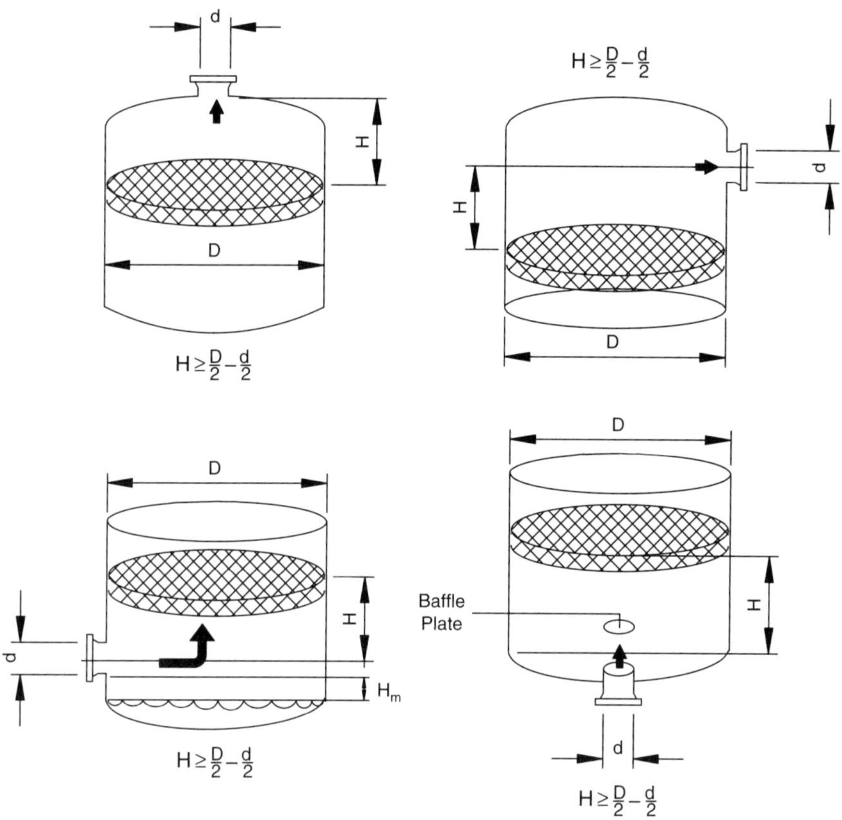

FIGURE 3.37. Dimensions for the placement of a wire-mesh mist extractor [H represents minimum height, and H_m must be at least 1 ft (305 mm).]

of a microfiber mist extractor can be 3–150 times that of a wire-mesh unit of equal volume.

There are two categories of these units, depending on whether droplet capture is via inertial impaction, interception, or Brownian diffusion. Only the diffusion type can remove droplets less than 2 μm. As with wire-mesh pads, microfiber units that operate in the inertial impaction mode have a minimum velocity below which efficiency drops off significantly. Microfiber units that operate in the diffusion mode have no such lower velocity limit. In fact, efficiency continues to improve as the gas velocity is reduced to zero.

For impaction-type microfiber units, the maximum velocity is usually set by the onset of re-entrainment, just as in the case of wire-mesh and vane devices. For microfiber units operating in the diffusion mode, the upper velocity can be set by re-entrainment, loss of efficiency, or pressure drop. Typical velocity ranges from 20 to

60 ft/min (60–180 m/min) for impaction-type units, compared to 1–4 ft/min (3–12 m/min) for units in the diffusion mode.

As with other mist extractors, each microfiber supplier has developed data on the capacity, pressure drop, and efficiency correlations for its products. Table 3.1 summarizes the major parameters that should be considered when selecting a mist extractor. For more detailed information, see Fabian et al. (1993).

Other Configurations

Some separators use centrifugal mist extractors, discussed earlier in this chapter, that cause liquid droplets to be separated by centrifugal force (Figures 3.38 and 3.39). These units can be more efficient than either wire-mesh or vanes and are the least susceptible to plugging. However, they are not in common use in production operations because their removal efficiencies are sensitive to small changes in flow. In addition, they require relatively large pressure drops to create the centrifugal force. To a lesser extent, random packing is sometimes used for mist extraction, as shown in Figure 3.40. The packing acts as a coalescer.

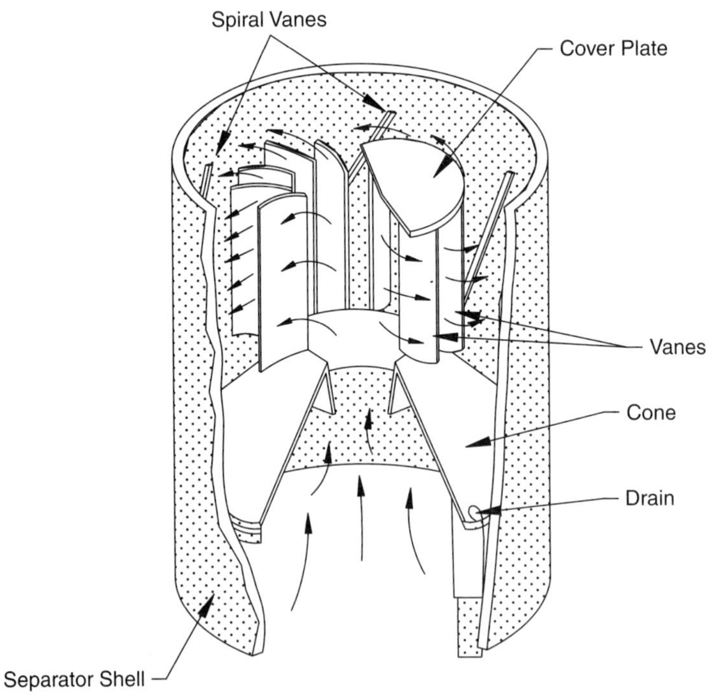

FIGURE 3.38. Centrifugal mist extractor.

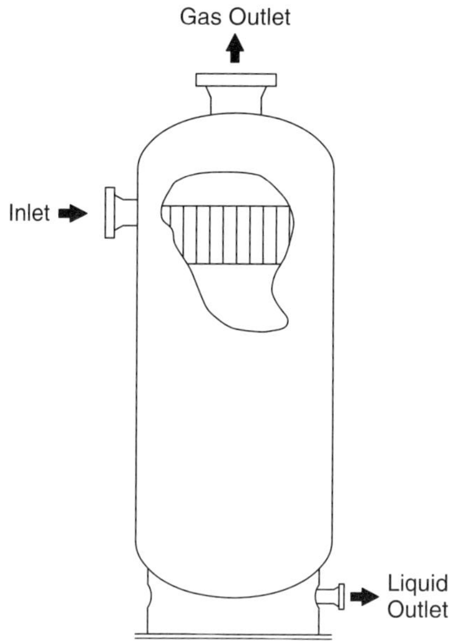

FIGURE 3.39. Vertical separator fitted with a centrifugal mist element (courtesy of Peerless Manufacturing Co.).

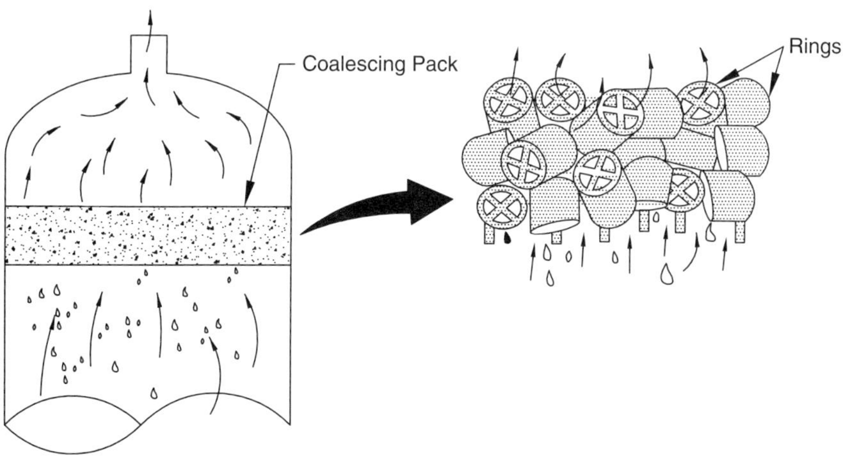

FIGURE 3.40. A coalescing pack mist extractor.

Final Selection

The selection of a type of mist extractor involves a typical cost-benefit analysis. Wire-mesh pads are the cheapest, but mesh pads are the most susceptible to plugging with paraffins, gas hydrates, and so forth. With age, mesh pads also tend to deteriorate and release wires and/or chunks of the pad into the gas stream. This can be extremely damaging to downstream equipment, such as compressors. Vane units, on the other hand, are more expensive. Typically, vane units are less susceptible to plugging and deterioration than mesh pads. Microfiber units are the most expensive and are capable of capturing very small droplets but, like wire mesh pads, are susceptible to plugging. The selection of a type of mist extractor is affected by the fluid characteristics, the system requirements, and the cost.

It is recommended that the sizing of mist extractors should be left to the manufacturer. Experience indicates that if the gravity settling section is designed to remove liquid droplets of 500 μm or smaller diameter, there will be sufficient space to install a mist extractor.

3.6 Potential Operating Problems

3.6.1 Foamy Crude

The major cause of foam in crude oil is the presence of impurities other than water, which are impractical to remove before the stream reaches the separator. One impurity that almost always causes foam is CO_2. Sometimes completion and workover fluids, that are incompatible with the wellbore fluids, may also cause foam. Foam presents no problem within a separator if the internal design ensures adequate time or sufficient coalescing surface for the foam to break.

Foaming in a separating vessel is a three-fold problem:

1. Mechanical control of liquid level is aggravated because any control device must deal with essentially three liquid phases instead of two.
2. Foam has a large volume-to-weight ratio. Therefore, it can occupy much of the vessel space that would otherwise be available in the liquid collecting or gravity settling sections.
3. In an uncontrolled foam bank, it becomes impossible to remove separated gas or degassed oil from the vessel without entraining some of the foamy material in either the liquid or gas outlets.

The foaming tendencies of any oil can be determined with laboratory tests. Only laboratory tests, run by qualified service companies, can qualitatively determine an oil's foaming tendency. One such test is ASTM D 892, which involves bubbling air through the oil.

Alternatively, the oil may be saturated with its associated gas and then expanded in a gas container.

This alternative test more closely models the actual separation process. Both of these tests are qualitative. There is no standard method of measuring the amount of foam produced or the difficulty in breaking the foam. Foaming is not possible to predict ahead of time without laboratory tests. However, foaming can be expected where CO_2 is present in small quantities (1–2%). It should be noted that the amount of foam is dependent on the pressure drop to which the inlet liquid is subjected, as well as the characteristics of the liquid at separator conditions.

Comparison of foaming tendencies of a known oil to a new one, about which no operational information is known, provides an understanding of the relative foam problem that may be expected with the new oil as weighed against the known oil. A related amount of adjustment can then be made in the design parameters, as compared to those found satisfactory for the known case.

The effects of temperature on a foamy oil are interesting. Changing the temperature at which a foamy oil is separated has two effects on the foam. The first effect is to change the oil viscosity. That is, an increase in temperature will decrease the oil viscosity, making it easier for the gas to escape from the oil. The second effect is to change the gas–oil equilibrium. A temperature increase will increase the amount of gas, which evolves from the oil.

It is very difficult to predict the effects of temperature on the foaming tendencies of an oil. However, some general observations have been made. For low API gravity crude (heavy oils) with low GORs, increasing the operating temperature decreases the oils' foaming tendencies. Similarly, for high API crude (light oils) with high GORs, increasing the operating temperature decreases the oils' foaming tendencies. However, increasing the operating temperature for a high-API gravity crude (light oil) with low GORs may increase the foaming tendencies. Oils in the last category are typically rich in intermediates, which have a tendency to evolve to the gas phase as the temperature increases. Accordingly, increasing the operating temperature significantly increases gas evolution, which in turn increases the foaming tendencies.

Foam depressant chemicals often will do a good job in increasing the capacity of a given separator. However, in sizing a separator to handle a specific crude, the use of an effective depressant should not be assumed because characteristics of the crude and of the foam may change during the life of the field. Also, the cost of foam depressants for high-rate production may be prohibitive. Sufficient capacity should be provided in the separator to handle the anticipated production without use of a foam depressant or inhibitor. Once placed in operation, a foam depressant may allow more throughput than the design capacity.

3.6.2 Paraffin

Separator operation can be adversely affected by an accumulation of paraffin. Coalescing plates in the liquid section and mesh pad mist extractors in the gas section are particularly prone to plugging by accumulations of paraffin. Where it is determined that paraffin is an actual or potential problem, the use of plate-type or centrifugal mist extractors should be considered.

Manways, handholes, and nozzles should be provided to allow steam, solvent, or other types of cleaning of the separator internals. The bulk temperature of the liquid should always be kept above the cloud point of the crude oil.

3.6.3 Sand

Sand can be very troublesome in separators by causing cutout of valve trim, plugging of separator internals, and accumulation in the bottom of the separator. Special hard trim can minimize the effects of sand on the valves. Accumulations of sand can be removed by periodically injecting water or steam in the bottom of the vessel so as to suspend the sand during draining. Figure 3.25 is a cutaway of a sand wash and drain system fitted into a horizontal separator fitted with sand jets and an inverted trough.

Sometimes a vertical separator is fitted with a cone bottom. This design would be used if sand production was anticipated to be a major problem. The cone is normally at an angle of between 45° and 60° to the horizontal. Produced sand may have a tendency to stick to steel at 45°. If a cone is installed, it could be part of the pressure-containing walls of the vessel (Figure 3.41), or for structural reasons, it could be installed internal to the vessel cylinder (Figure 3.42). In such a case, a gas equalizing line must be installed to assure that the vapor behind the cone is always in pressure equilibrium with the vapor space.

Plugging of the separator internals is a problem that must be considered in the design of the separator. A design that will promote good separation and have a minimum of traps for sand accumulation may be difficult to attain, since the design that provides the best mechanism for separating the gas, oil, and water phases probably will also provide areas for sand accumulation. A practical balance for these factors is the best solution.

3.6.4 Liquid Carryover

Liquid carryover occurs when free liquid escapes with the gas phase and can indicate high liquid level, damage to vessel internals, foam, improper design, plugged liquid outlets, or a flow rate that exceeds the vessel's design rate. Liquid carryover can usually be prevented by

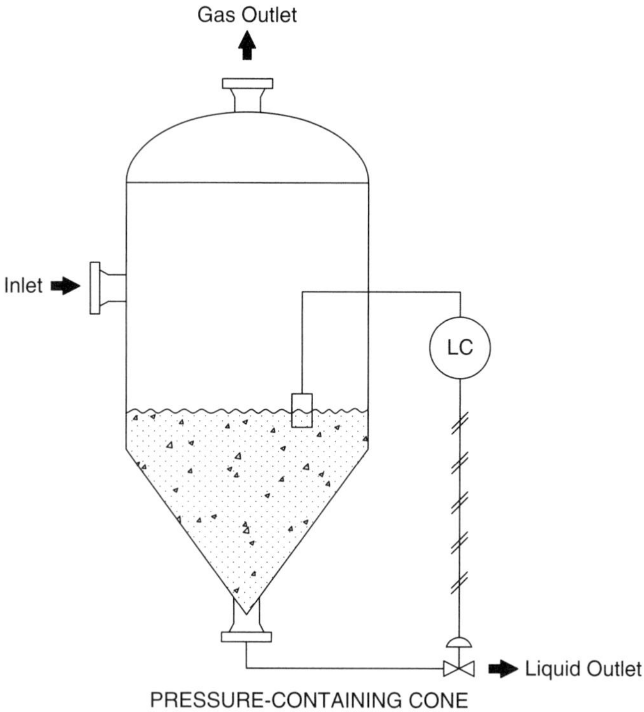

FIGURE 3.41. Vertical separator with a pressure-containing cone bottom used to collect solids.

installing a level safety high (LSH) sensor that shuts in the inlet flow to the separator when the liquid level exceeds the normal maximum liquid level by some percentage, usually 10–15%.

3.6.5 Gas Blowby

Gas blowby occurs when free gas escapes with the liquid phase and can be an indication of low liquid level, vortexing, or level control failure. This could lead to a very dangerous situation. If there is a level control failure and the liquid dump valve is open, the gas entering the vessel will exit the liquid outlet line and would have to be handled by the next downstream vessel in the process. Unless the downstream vessel is designed for the gas blowby condition, it can be overpressured. Gas blowby can usually be prevented by installing a level safety low sensor (LSL) that shuts in the inflow and/or outflow to the vessel when the liquid level drops to 10–15% below the lowest operating level. In addition, downstream process components should be equipped with a pressure safety high (PSH) sensor and a pressure safety valve (PSV) sized for gas blowby.

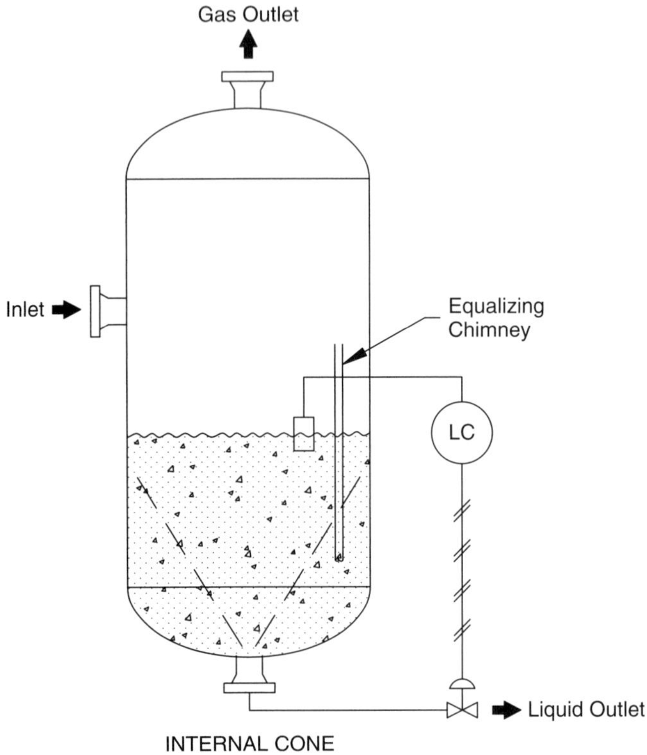

FIGURE 3.42. Vertical separator fitted with an internal cone bottom and an equalizing line.

3.6.6 Liquid Slugs

Two-phase flow lines and pipelines tend to accumulate liquids in low spots in the lines. When the level of liquid in these low spots rises high enough to block the gas flow, then the gas will push the liquid along the line as a slug. Depending on the flow rates, flow properties, length and diameter of the flow line, and the elevation change involved, these liquid slugs may contain large liquid volumes.

Situations in which liquid slugs may occur should be identified prior to the design of a separator. The normal operating level and the high-level shutdown on the vessel must be spaced far enough apart to accommodate the anticipated slug volume. If sufficient vessel volume is not provided, then the liquid slugs will trip the high-level shutdown.

When liquid slugs are anticipated, slug volume for design purposes must be established. Then the separator may be sized for liquid flow-rate capacity using the normal operating level. The location of the high-level set point may be established to provide the slug volume between the normal level and the high level. The separator size must then be checked to ensure that sufficient gas capacity is provided even when the liquid is at the high-level set point. This check of gas capacity is particularly important for horizontal separators because, as the liquid level rises, the gas capacity is decreased. For vertical separators, sizing is easier, as sufficient height for the slug volume may be added to the vessel's seam-to-seam length.

Often the potential size of the slug is so great that it is beneficial to install a large pipe volume upstream of the separator. The geometry of these pipes is such that they operate normally empty of liquid, but fill with liquid when the slug enters the system. This is the most common type of slug catcher used when two-phase pipelines are routinely pigged. Figure 3.15 is a schematic of a liquid finger slug catcher.

3.7 Design Theory

In the gravity settling section of a separator, liquid droplets are removed using the force of gravity. Liquid droplets, contained in the gas, settle at a terminal or "settling" velocity. At this velocity, the force of gravity on the droplet or "negative buoyant force" equals the drag force exerted on the droplet due to its movement through the continuous gas phase. The drag force on a droplet may be determined from the following equation:

$$F_{\mathrm{D}} = C_{\mathrm{D}} A_{\mathrm{d}} \rho (V^2/2g) \tag{3.3}$$

where

F_{D} = drag force, $\mathrm{lb_f}$ (N),
C_{D} = drag coefficient,
A_{d} = cross-sectional area of the droplet, ft^2 (m^2),
ρ = density of the continuous phase, $\mathrm{lb/ft}^3$ ($\mathrm{kg/m}^3$),
V_{t} = terminal (settling velocity) of the droplet, ft/sec (m/sec),
g = gravitational constant, 32.2 $\mathrm{lb_m ft/lb_f\ sec^2}$ ($\mathrm{m/sec^2}$).

If the flow around the droplet were laminar, then Stokes' law would govern and

$$C_{\mathrm{D}} = \frac{24}{Re} \tag{3.4}$$

where Re = Reynolds number, which is dimensionless.

It can be shown that in such a gas the droplet settling velocity would be given by:

Field units

$$V_t = \frac{1.78 \times 10^{-6}(\Delta SG)d_m^2}{\mu} \tag{3.5a}$$

SI units

$$V_t = \frac{5.56 \times 10^{-7}(\Delta SG)d_m^2}{\mu}, \tag{3.5b}$$

where

ΔSG = difference in specific gravity relative to water of the droplet and the gas,
d_m = droplet diameter, μm,
μ = viscosity of the gas, cp.

Unfortunately, for production facility designs it can be shown that Stokes' law does not govern, and the following more complete formula for drag coefficient must be used (refer to Figure 3.43):

$$C_D = \frac{24}{Re} + \frac{3}{Re^{1/2}} + 0.34 \tag{3.6}$$

Equating drag and buoyant forces, the terminal settling velocity is given by

Field units

$$V_t = 0.0119\left[\left(\frac{\rho_l - \rho_g}{\rho_g}\right)\frac{d_m}{C_D}\right]^{1/2} \tag{3.7a}$$

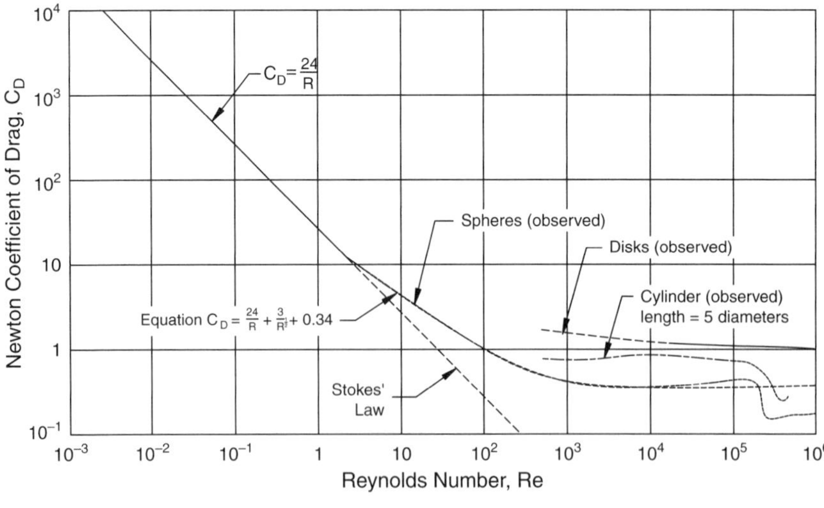

FIGURE 3.43. Coefficient of drag for varying magnitudes of Reynolds number.

SI units

$$V_t = 0.0036\left[\left(\frac{\rho_l - \rho_g}{\rho_g}\right)\frac{d_m}{C_D}\right]^{1/2} \quad (3.7b)$$

where

ρ_l = density of liquid, lb/ft^3 (kg/m^3),

ρ_g = density of the gas at the temperature and pressure in the separator, lb/ft^3 (kg/m^3).

Equations (3.7a) and (3.7b) are derived as follows:

C_D = constant.

For $C_D = 0.34$, Field units:

$$V_t = 0.0204\left[\left(\frac{\rho_l - \rho_g}{\rho_g}\right)d_m\right]^{1/2}.$$

For $C_D = 0.34$, SI units:

$$V_t = 0.0062\left[\left(\frac{\rho_l - \rho_g}{\rho_g}\right)d_m\right]^{1/2}.$$

Equations (3.6) and (3.7) can be solved by an iterative process. Start by assuming a value of C_D, such as 0.34, and solve Equation (3.7) for V_t. Then, using V_t, solve for Re. Then, Equation (3.6) may be solved for C_D. If the calculated value of C_D equals the assumed value, the solution has been reached. If not, then the procedure should be repeated using the calculated C_D as a new assumption. The original assumption of 0.34 for C_D was used because this is the limiting value for large Reynolds numbers. The iterative steps are shown below:

Field units

1. Start with

$$V_t = 0.0204\left[\frac{(\rho_l - \rho_g)}{\rho_g}d_m\right]^{1/2}.$$

2. Calculate

$$Re = 0.0049\frac{\rho_g d_m V}{\mu}.$$

3. From Re, calculate C_D using

$$C_D = \frac{24}{\text{Re}} + \frac{3}{\text{Re}^{1/2}} + 0.34.$$

4. Recalculate V_t using

$$V_t = 0.0119\left[\frac{(\rho_l - \rho_g)}{\rho_g}\frac{d_m}{C_D}\right]^{1/2}.$$

5. Go to step 2 and iterate.

SI units

1. Start with

$$V_1 = 0.0062\left[\frac{(\rho_l - \rho_g)d_m}{\rho_g}\right]^{1/2}.$$

2. Calculate

$$\mathrm{Re} = 0.001\frac{\rho_g d_m V}{\mu}.$$

3. From Re, calculate C_D using

$$C_D = \frac{24}{\mathrm{Re}} + \frac{3}{\mathrm{Re}^{1/2}} + 0.34.$$

4. Recalculate V_t using

$$V_t = 0.0036\left[\frac{(\rho_l - \rho_g)}{\rho_g}\frac{d_m}{C_D}\right]^{1/2}.$$

5. Go to step 2 and iterate.

3.7.1 Droplet Size

The purpose of the gravity settling section of the vessel is to condition the gas for final polishing by the mist extractor. To apply the settling equations to separator sizing, a liquid droplet size to be removed must be selected. From field experience, it appears that if 140-μm droplets are removed in this section, the mist extractor will not become flooded and will be able to perform its job of removing those droplets between 10- and 140-μm diameters. The gas capacity design equations in this section are all based on 140-μm removal. In some cases, this will give an overly conservative solution. The techniques used here can be easily modified for any droplet size.

In this book we are addressing separators used in oil field facilities. These vessels usually require a gravity settling section. There are special cases where the separator is designed to remove only very small quantities of liquid that could condense due to temperature or pressure changes in a stream of gas that has already passed through a separator and a mist extractor. These separators, commonly called gas scrubbers, could be designed for removal of droplets on the order of 500 μm without fear of flooding their mist extractors. Fuel gas scrubbers, compressor suction scrubbers, and contact tower inlet scrubbers are examples of vessels to which this might apply.

Flare or vent scrubbers are designed to keep large slugs of liquid from entering the atmosphere through the vent or relief systems. In vent systems the gas is discharged directly to the atmosphere, and it is common to design the scrubbers for removal of 300- to 500-μm droplets in the gravity settling section. A mist extractor is not included because of the possibility that it might get plugged, thus creating a safety hazard.

In flare systems, where the gas is discharged through a flame, there is the possibility that burning liquid droplets could fall to the ground before being consumed. It is still common to size the gravity settling section for 300- to 500-μm removal, which the API guideline for refinery flares indicates is adequate to ensure against a falling flame. In critical locations, such as offshore platforms, many operators include a mist extractor as an extra precaution against a falling flame.

If a mist extractor is used, it is necessary to provide safety relief protection around the mist extractor in the event that it becomes plugged.

3.7.2 Retention Time

To ensure that the liquid and gas reach equilibrium at separator pressure, a certain liquid storage is required. This is defined as "retention time" or the average time a molecule of liquid is retained in the vessel, assuming plug flow. The retention time is thus the volume of the liquid storage in the vessel divided by the liquid flow rate.

For most applications retention times between 30 sec and 3 min have been found to be sufficient. Where foaming crude is present, retention times up to four times this amount may be needed. In the absence of liquid or laboratory data, the guidelines presented in Table 3.2 can be used.

TABLE 3.2
Retention time for two-phase separators

°API Gravity	*Retention Time (min)*
35+	0.5–1
30	2
25	3
20−	4+

If foam exists, increase above retention times by a factor of 2–4.
If high CO_2 exists, use a minimum of 5-min retention time.

3.7.3 Liquid re-entrainment

Liquid re-entrainment is a phenomenon caused by high gas velocity at the gas–liquid interface of a separator. Momentum transfer from the gas to the liquid causes waves and ripples in the liquid, and then droplets are broken away from the liquid phase.

The general rule of thumb that calls for limiting the slenderness ratio to a maximum of 4 or 5 is applicable for half-full horizontal separators. Liquid re-entrainment should be particularly considered for high-pressure separators sized on gas-capacity constraints. It is more likely at higher operating pressures (>1000 psig or >7000 kPa) and higher oil viscosities (<30 °API). For more specific limits, see Viles (1993).

3.8 Separator Design

3.8.1 Horizontal Separators Sizing—Half Full

The guidelines presented in this section can be used for the initial sizing of a horizontal separator 50% full of liquid. They are meant to complement, and not replace, operating experience. Determination of the type and size of separator must be on an individual basis. All the functions and requirements should be considered, including the uncertainties in design flow rates and fluid properties. For this reason, there is no substitute for good engineering evaluations of each separator by the design engineer. The trade-off between design size and details and uncertainties in design parameters should not be left to manufacturer recommendations or rule of thumb.

When sizing a horizontal separator, it is necessary to choose a seam-to-seam vessel length and a diameter. This choice must satisfy the conditions for gas capacity that allow the liquid droplets to fall from the gas to the liquid volume as the gas traverses the effective length of the vessel. It must also provide sufficient retention time to allow the liquid to reach equilibrium. Figure 3.44 shows a vessel 50% full of liquid, which is the model used to develop sizing equations for a horizontal separator.

3.8.2 Gas Capacity Constraint

The principles of liquid droplets settling through a gas can be used to develop an equation to size a separator for a gas flow rate. The gas capacity constraint equations are based on setting the gas retention time equal to the time required for a droplet to settle to the liquid interface. For a vessel 50% full of liquid, and separation of 100-μm liquid droplets from the gas, the following equation may be derived:

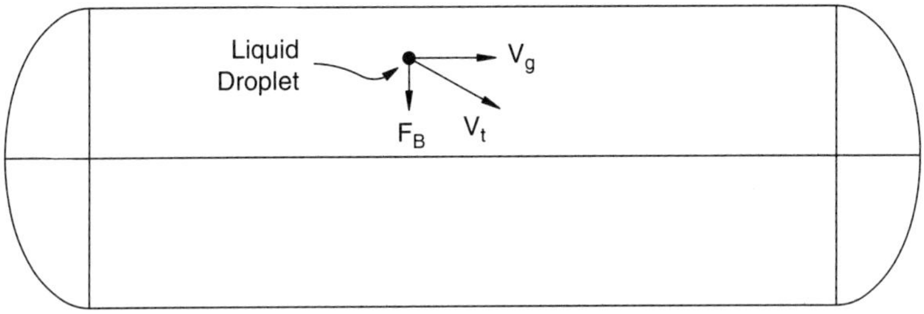

FIGURE 3.44. Model of a horizontal separator.

Field units

$$dL_{\text{eff}} = 420\left[\frac{TZQ_{\text{g}}}{P}\right]\left[\left(\frac{\rho_{\text{g}}}{\rho_{\text{l}}-\rho_{\text{g}}}\right)\frac{C_{\text{D}}}{d_{\text{m}}}\right]^{1/2} \tag{3.8a}$$

SI units

$$dL_{\text{eff}} = 34.5\left[\frac{TZQ_{\text{g}}}{P}\right]\left[\left(\frac{\rho_{\text{g}}}{\rho_{\text{l}}-\rho_{\text{g}}}\right)\frac{C_{\text{D}}}{d_{\text{m}}}\right]^{1/2} \tag{3.8b}$$

where

d = vessel internal diameter, in. (mm),
L_{eff} = effective length of the vessel where separation occurs, ft (m),
T = operating temperature, °R (°K),
Q_{g} = gas flow rate, MMscfd (scmh),
P = operating pressure, psia (kPa),
Z = gas compressibility,
C_{D} = drag coefficient,
d_{m} = liquid droplet to be separated, micron,
ρ_{g} = density of gas, lb/ft^3 (kg/m^3),
ρ_{l} = density of liquid, lb/ft^3(kg/m^3).

3.8.3 Liquid Capacity Constraint

Two-phase separators must be sized to provide some liquid retention time so the liquid can reach phase equilibrium with the gas. For a vessel 50% full of liquid, with a specified liquid flow rate and retention time, the following may be used to determine vessel size.

Field units

$$d^2 L_{\text{eff}} = \frac{t_r Q_l}{0.7} \tag{3.9a}$$

SI units

$$d^2 L_{\text{eff}} = 42,441 t_r Q_l \tag{3.9b}$$

where

t_r = desired retention time for the liquid, min,
Q_l = liquid flow rate, bpd (m^3/h).

3.8.4 Seam-to-Seam Length

The effective length may be calculated from Equations (3.8a)–(3.9b). From this, a vessel seam-to-seam length may be determined. The actual required seam-to-seam length is dependent on the physical design of the internals of the vessel.

As shown in Figure 3.45, for vessels sized on a gas capacity basis, some portion of the vessel length is required to distribute the flow evenly near the inlet diverter. Another portion of the vessel length is required for the mist extractor. The length of the vessel between the inlet diverter and the mist extractor with evenly distributed flow is

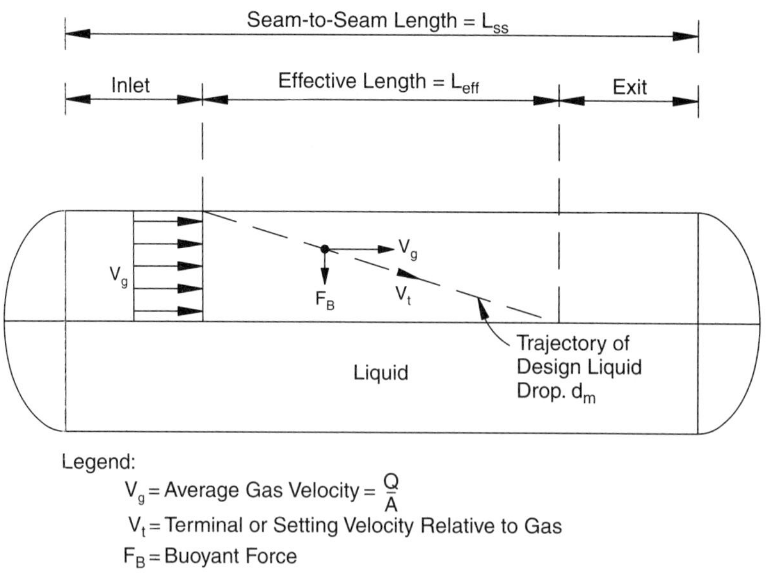

FIGURE 3.45. Approximate seam-to-seam length of a horizontal separator one-half full.

the L_{eff} calculated from Equations (3.8a) and (3.8b). As a vessel's diameter increases, more length is required to evenly distribute the gas flow. However, no matter how small the diameter may be, a portion of the length is still required for the mist extractor and flow distribution. Based on these concepts coupled with field experience, the seam-to-seam length of a vessel may be estimated as the larger of the following.

Field units

$$L_{ss} = L_{eff} + \frac{d}{12} \text{ for gas capacity} \tag{3.10a}$$

SI units

$$L_{ss} = L_{eff} + \frac{d}{1000} \text{ for gas capacity} \tag{3.10b}$$

For vessels sized on a liquid capacity basis, some portion of the vessel length is required for inlet diverter flow distribution and liquid outlet. The seam-to-seam length should not exceed the following:

$$L_{ss} = (4/3)L_{eff}. \tag{3.11}$$

3.8.5 Slenderness Ratio

Equations (3.8a)–(3.9b) allow for various choices of diameter and length. For each vessel design, a combination of L_{eff} and d exists that will minimize the cost of the vessel. It can be shown that the smaller the diameter, the less the vessel will weigh and thus the lower its cost. There is a point, however, where decreasing the diameter increases the possibility that high velocity in the gas flow will create waves and re-entrain liquids at the gas–liquid interface.

Experience has shown that if the gas capacity governs and the length divided by the diameter, referred to as the "slenderness ratio," is greater than 4 or 5, re-entrainment could become a problem. Equation (3.11) indicates that slenderness ratios must be at least 1 or more. Most two-phase separators are designed for slenderness ratios between 3 and 4. Slenderness ratios outside the 3–4 range may be used, but the design should be checked to assure that re-entrainment will not occur.

3.8.6 Procedure for Sizing Horizontal Separators—Half Full

1. The first step in sizing a horizontal separator is to establish the design basis. This includes specifying the maximum and minimum flow rates, operating pressure and temperature, droplet size to be removed, etc.

2. Prepare a table with calculated values of L_{eff} for selected values of d that satisfy Equations (3.8a) and (3.8b), and the gas capacity constraint. Calculate L_{ss} using Equations (3.10a) and (3.10b).

   Field units

$$L_{eff}d = 420\frac{TZQ_g}{P}\left[\left(\frac{\rho_g}{\rho_l - \rho_g}\right)\frac{C_D}{d_m}\right]^{1/2}$$

   SI units

$$L_{eff}d = 34.5\frac{TZQ_g}{P}\left[\left(\frac{\rho_g}{\rho_l - \rho_g}\right)\frac{C_D}{d_m}\right]^{1/2}$$

3. For the same values of d, calculate values of L_{eff} using Equations (3.9a) and (3.9b) for liquid capacity and list these values in the same table. Calculate L_{ss} using Equation (3.11).

   Field units

$$d^2L_{eff} = \frac{t_rQ_l}{0.7}$$

   SI units

$$d^2L_{eff} = 42,441t_rQ_l$$

4. For each d, the larger L_{eff} should be used.
5. Calculate the slenderness ratio, $12L_{eff}/<d_o(1000L_{eff}/<d_o)$, and list for each d. Select a combination of d and L_{ss} that has a slenderness ratio between 3 and 4. Lower ratios can be chosen if dictated by available space, but they will probably be more expensive. Higher ratios can be chosen if the vessel is checked for re-entrainment.
6. When making a final selection, it is always more economical to select a standard vessel size. Vessels with outside diameters up through 24 in. (600 mm) have nominal pipe dimensions. Vessels with outside diameters larger than 24 in. (600 mm) are typically rolled from plate with diameter increments of 6 in. (150 mm). The shell seam-to-seam length is expanded in 2.5-ft (750-mm) segments and is usually from 5 ft to 10 ft (1500–3000 mm). Standard separator vessel sizes may be obtained from API 12J.

3.8.7 Horizontal Separators Sizing Other Than Half Full

The majority of oil field two-phase separators are designed with the liquid level at the vessel centerline—that is, 50% full of liquid. For a vessel other than 50% full of liquid, Equations (3.12a)–(3.13b) apply.

These equations were derived using the actual gas and liquid areas to calculate gas velocity and liquid volume (Figure 3.46).
Gas capacity constraint

Field units

$$dL_{\text{eff}} = 420\left(\frac{1-\beta}{1-\alpha}\right)\left[\frac{TZQ_{\text{g}}}{P}\right]\left[\left(\frac{\rho_{\text{g}}}{\rho_{\text{l}}-\rho_{\text{g}}}\right)\frac{C_{\text{D}}}{d_{\text{m}}}\right]^{1/2}, \tag{3.12a}$$

where

$$\left(\frac{1-\beta}{1-\alpha}\right) = \text{design constant (Figure 3.47)}.$$

SI units

$$dL_{\text{eff}} = 34.5\left(\frac{1-\beta}{1-\alpha}\right)\left[\frac{TZQ_{\text{g}}}{P}\right]\left[\left(\frac{\rho_{\text{g}}}{\rho_{\text{l}}-\rho_{\text{g}}}\right)\frac{C_{\text{D}}}{d_{\text{m}}}\right]^{1/2}, \tag{3.12b}$$

where

$$\left(\frac{1-\beta}{1-\alpha}\right) = \text{design constant (Figure 3.47)}.$$

Liquid capacity constraint

Field units

$$d^2 L_{\text{eff}} = \frac{t_{\text{r}} Q_{\text{l}}}{1.4\alpha}, \tag{3.13a}$$

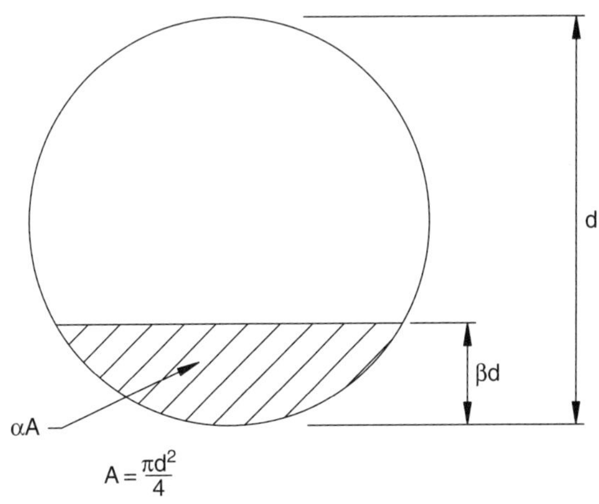

FIGURE 3.46. Definition of parallel areas.

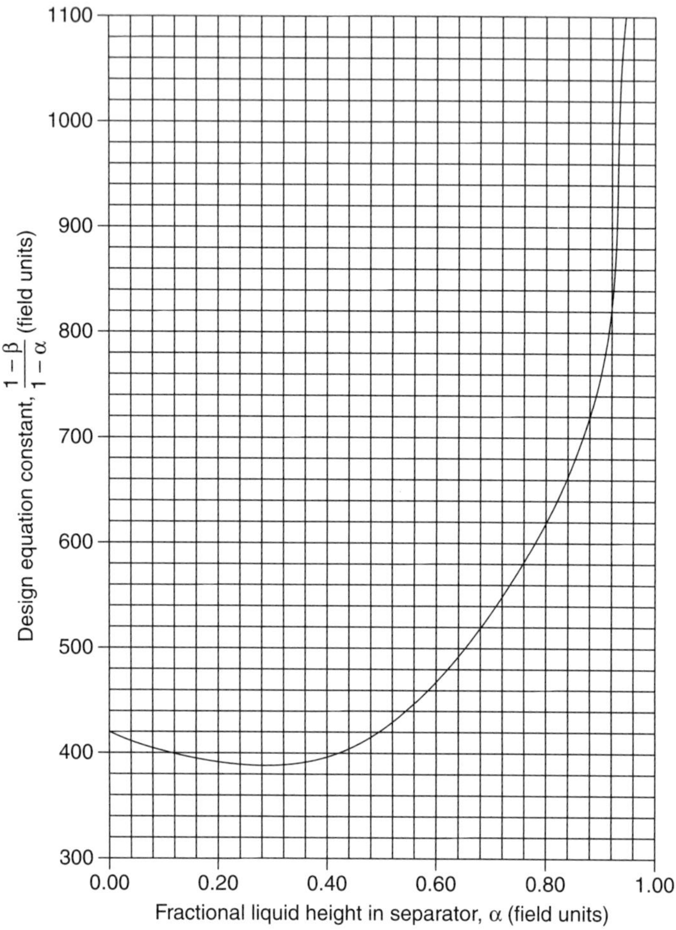

FIGURE 3.47. Gas capacity constraint design constant versus liquid height of a cylinder for a horizontal separator other than 50% full of liquid.

where

α = design constant

If β is known, α can be determined from Figure 3.48.

SI units

$$d^2 L_{\text{eff}} = \frac{21,221 t_r Q_l}{\alpha}, \tag{3.13b}$$

where

α = design constant

If β is known, α can be determined from Figure 3.48.

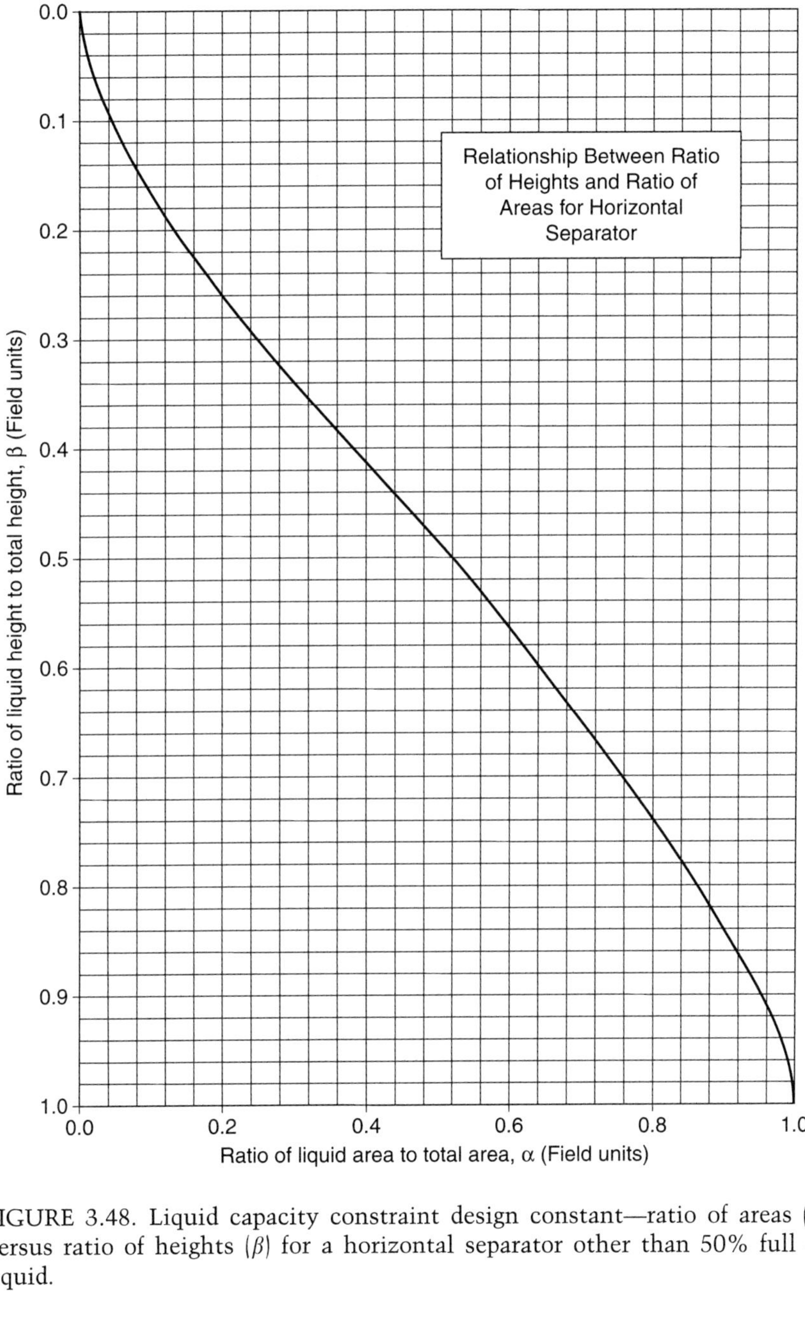

FIGURE 3.48. Liquid capacity constraint design constant—ratio of areas (α) versus ratio of heights (β) for a horizontal separator other than 50% full of liquid.

Vertical Separators' Sizing

The guidelines presented in this section can be used for initial sizing of a vertical two-phase separator. They are meant to complement, and not replace, operating experience. Determination of the type and size of separator must be on an individual basis. All the functions and requirements should be considered, including the uncertainties in design flow rates and properties. For this reason, there is no substitute for good engineering evaluations of each separator by the design engineer. The trade-off between design size and details and uncertainties should not be left to manufacturer recommendations or rules of thumb.

In vertical separators, a minimum diameter must be maintained to allow liquid droplets to separate from the vertically moving gas. The liquid retention time requirement specifies a combination of diameter and liquid volume height. Any diameter greater than the minimum required for gas capacity can be chosen. Figure 3.49 shows the model used for a vertical separator.

Gas Capacity Constraint

The principles of liquid droplets settling through a gas can be used to develop an equation to size a separator for a gas flow rate. By setting the gas retention time equal to the time required for a droplet to settle to the liquid interface, the following equation may be derived.

Field units

$$d^2 = 5040\left[\frac{TZQ_g}{P}\right]\left[\left(\frac{\rho_g}{\rho_l - \rho_g}\right)\frac{C_D}{d_m}\right]^{1/2} \tag{3.14a}$$

SI units

$$d^2 = 34,444\left[\frac{TZQ_g}{P}\right]\left[\left(\frac{\rho_g}{\rho_l - \rho_g}\right)\frac{C_D}{d_m}\right]^{1/2} \tag{3.14b}$$

3.8.8 Liquid Capacity Constraint

Two-phase separators must be sized to provide some liquid retention time so the liquid can reach phase equilibrium with the gas. For a specified liquid flow rate and retention time, the following may be used to determine a vessel size.

Field units

$$d^2h = \frac{t_r Q_l}{0.12} \tag{3.15a}$$

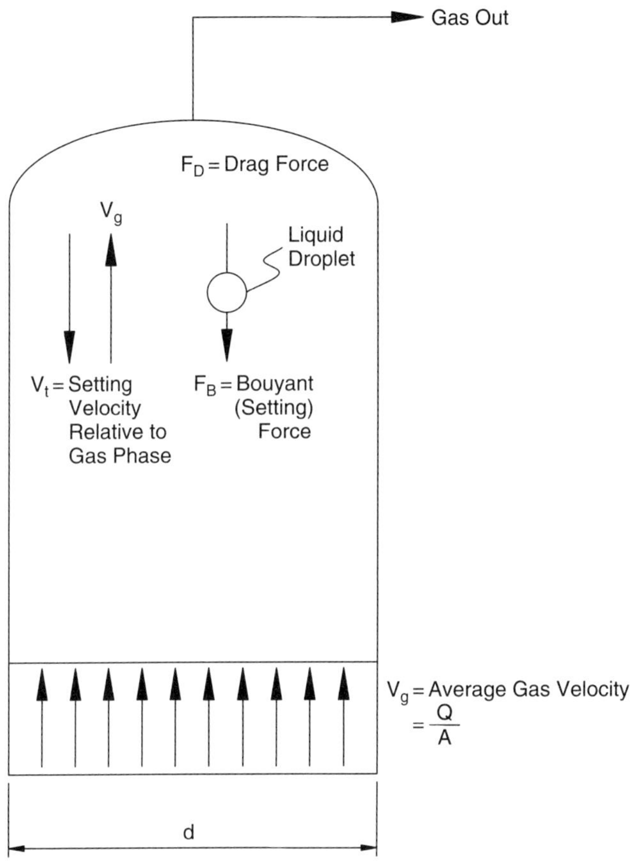

FIGURE 3.49. Model of a vertical separator.

SI units

$$d^2h = \frac{t_r Q_l}{4.713 \times 10^{-8}}, \qquad (3.15b)$$

where h = height of the liquid volume, in. (mm).

3.8.9 Seam-to-Seam Length

As with horizontal separators, the specific design of the vessel internals will affect the seam-to-seam length. The seam-to-seam length of vertical vessels may be estimated based on the diameter and liquid height. As shown in Figure 3.50, allowance must be made for the gas separation section and mist extractor and for any space below the water outlet. For screening purposes, the following may be used to estimate L_{ss}.

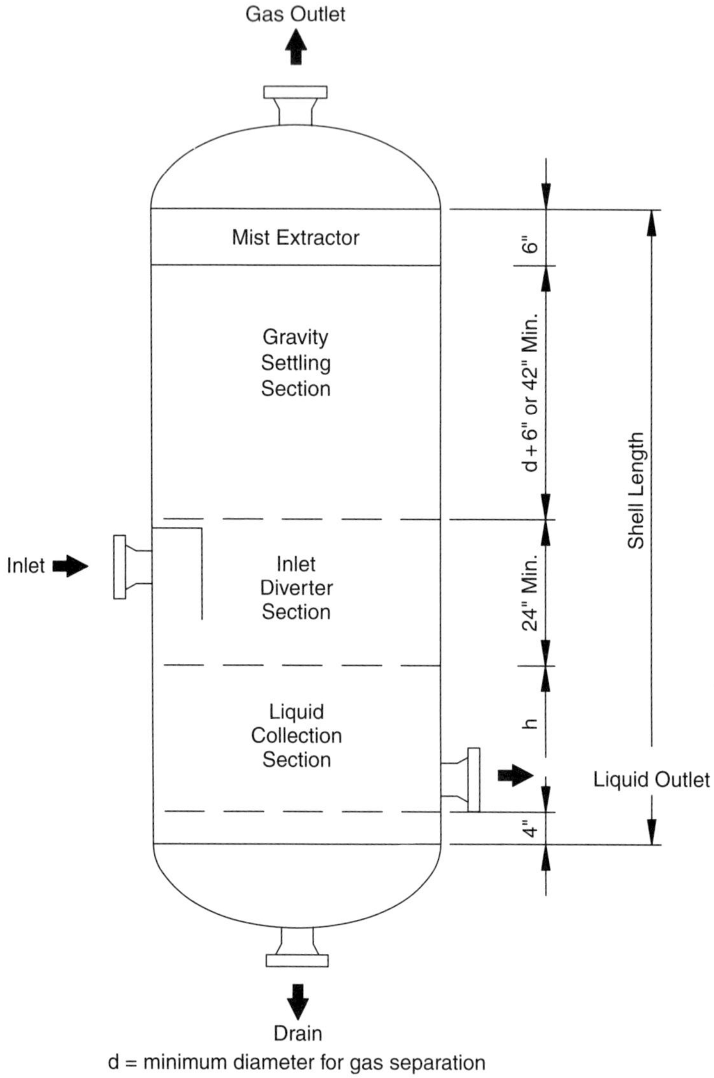

FIGURE 3.50. Approximate seam–seam shell length for a vertical separator.

Field units

$$L_{ss} = \frac{h + 76}{12} \text{ (for diameters} \leq 36 \text{ in.)} \tag{3.16a}$$

SI units

$$L_{ss} = \frac{h + 1930}{1000} \text{ (for diameters} \geq 194 \text{ mm)} \tag{3.16b}$$

Field units

$$L_{ss} = \frac{h + d + 40}{12} \text{ (for diameters} > 36 \text{ in.)} \tag{3.17a}$$

SI units

$$\frac{h + d + 1016}{1000} \text{ (for diameters} > 194 \text{ mm),} \tag{3.17b}$$

where

h = height of liquid level, in. (mm),
d = vessel internal diameter, in. (mm).

The larger of the L_{ss} values from Equations (3.16a)–(3.17b) should be used.

3.8.10 Slenderness Ratio

As with horizontal separators, the larger the slenderness ratio, the less expensive the vessel will be. In vertical separators whose sizing is liquid dominated, it is common to choose slenderness ratios no greater than 4 to keep the height of the liquid collection section to a reasonable level. Choices of between 3 and 4 are common, although height restrictions may force the choice of a lower slenderness ratio.

3.8.11 Procedure for Sizing Vertical Separators

1. The first step in sizing a vertical separator is to establish the design basis. This includes specifying the maximum and minimum flow rates, operating pressure and temperature, droplet size to be removed, and so on.
2. Equations (3.14a) and (3.14b) may be used to determine the minimum required d. Any diameter larger than this value may be used.
3. For a selected d, Equations (3.15a) and (3.15b) may be used to determine h.
4. From d and h, the seam-to-seam length may be estimated using Equations (3.16a)–(3.17b). The larger value of L_{ss} should be used.
5. Check the slenderness ratio to determine if it is less than 4.
6. When making a final selection, it is always more economical to select a standard vessel size. Vessels with outside diameters up through 24 in. (600 mm) have nominal pipe dimensions. Vessels with outside diameters larger than 24 in. (600 mm) are rolled from plate with diameter increments of 6 in. (150 mm). The shell seam-to-seam length is expanded in 2.5-ft (750-mm) segments and is usually from 5 ft to 10 ft (1500 mm–3000 mm). Standard separator vessel sizes may be obtained from API 12J.

3.8.12 Examples

Example 3.1: Sizing a Vertical Separator (Field Units)

Given:

Gas flow rate: 10 MMscfd at 0.6 specific gravity
Oil flow rate: 2000 BOPD at 40°API
Operating pressure: 1000 psia
Operating temperature: 60°F
Droplet size removal: 140 μm
Retention time: 3 min

Solution:

1. Calculate C_D

$$\rho_l = 62.4\left[\frac{141.5}{131.5 + 40}\right] = 51.5\frac{\text{lb}}{\text{ft}^3},$$

$$\rho_g = 2.70\frac{SP}{TZ}, \quad Z = 0.84 \text{ (from Chapter 1)},$$

$$\rho_g = 2.70\frac{(0.6)(1000)}{(520)(0.84)} = 3.71\text{lb/ft}^3,$$

$$d_m = 140\mu\text{m}, \quad \mu = 0.013 \text{ cp (from Chapter 1)}$$

Assume $C_D = 0.34$.

$$V_t = 0.0119\left[\left(\frac{51.5 - 3.71}{3.71}\right)\frac{140}{0.34}\right]^{1/2},$$

$$V_t = 0.867\text{ ft/sec}, \quad Re = 0.0049\left[\frac{(3.71)(140)(0.866)}{0.013}\right] = 16$$

$$C_D = \frac{24}{169.54} + \frac{3}{(169.54)^{1/2}} + 0.34, \quad C_D = 0.712.$$

Repeat using $C_D = 0.712$.

$$V_t = 0.599\text{ ft/sec}, \quad Re = 117, \quad C_D = 0.822.$$

Repeat:

$$V_t = 0.556, \quad Re = 110, \quad C_D = 0.844.$$

Repeat:

$$V_t = 0.548, \quad Re = 108, \quad C_D = 0.851.$$

Repeat:

$$V_t = 0.545, \quad Re = 108, \quad C_D = 0.854\text{—OK}.$$

2. Gas capacity constraint

$$d^2 = 5040\left[\frac{TZQ_g}{P}\right]\left[\left(\frac{\rho_g}{\rho_l - \rho_g}\right)\frac{C_D}{d_m}\right]^{1/2}, \quad Z = 0.84 \text{ (from Chapter 1)},$$

$$d^2 = 5040\left[\frac{(520)(0.84)(10)}{1,000}\right]\left[\left(\frac{3.71}{51.5 - 3.71}\right)\frac{0.851}{140}\right]^{1/2}, \quad d = 21.9\text{m}$$

3. Liquid capacity constraint

$$d^2 h = \frac{t_r Q_l}{0.12}$$

4. Compute combinations of d and h for various t_r (Table 3.3).
5. Compute seam-to-seam length (Table 3.3).

$$L_{ss} = \frac{h + 76}{12} \quad \text{or} \quad L_{ss} = \frac{h + d + 40}{12},$$

where d is the minimum diameter for gas capacity

6. Compute slenderness ratio: $12L_{ss}/d$. Choices in the range of 3–4 are most common (Table 3.3).
7. Choose a reasonable size with a diameter greater than that determined by the gas capacity. A 36-in. diameter by 10-ft. seam-to-seam separator provides slightly more than 3 min retention time with a diameter greater than 21.8 in. and a slenderness ratio of 3.2.

TABLE 3.3
Vertical separator example diameter versus length for liquid capacity constraint

t_r (min)	d (in.)	h (in.)	L_{ss} (ft.)	$S_R\left(\frac{12L_{ss}}{d}\right)$
3	24	86.8	13.6	6.8
	30	55.6	11.0	4.4
	36	38.6	9.6	3.2
	42	28.3	8.7	2.5
	48	21.7	8.1	2.0
2	24	57.9	11.2	5.6
	30	37.0	9.4	3.8
	36	25.7	8.5	2.8
	42	18.9	7.9	2.3
1	24	28.9	8.7	4.4
	30	18.5	7.9	3.2
	36	12.9	7.4	2.5

Example 3.2: Sizing a Horizontal Separator (field units)

Given:

Gas flow rate: 10 MMscfd at 0.6 specific gravity
Oil flow rate: 2000 BOPD at 40 °API
Operating pressure: 1000 psia
Operating temperature: 60 °F
Droplet size removal: 140 μm
Retention time: 3 min

Solution:

1. Calculate C_D (same as Example 3.1).

$$C_D = 0.851$$

2. Gas capacity constraint

$$dL_{eff} = 420\left[\frac{TZQ_g}{P}\right]\left[\left(\frac{\rho_g}{\rho_l - \rho_g}\right)\frac{C_D}{d_m}\right]^{1/2},$$

$Z = 0.84$ (from Chapter 1),

$$dL_{eff} = 420\left[\frac{(520)(0.84)(10)}{1000}\right]\left[\left(\frac{3.71}{51.5 - 3.71}\right)\frac{0.851}{140}\right]^{1/2} = 55.04.$$

3. Liquid capacity constraint

$$d^2 L_{eff} = \frac{t_r Q_l}{0.7}$$

4. Compute combinations of d and L_{ss} for gas and liquid capacity.
5. Compute seam-to-seam length for various d (Table 3.4).

$$L_{ss} = L_{eff} + \frac{d}{12}$$

TABLE 3.4
Horizontal separator example diameter versus length

d (ft)	*Gas* L_{eff} (ft)	*Liquid* L_{eff} (ft)	L_{ss} (ft)	$12L_{ss}/d$
16	2.5	33.5	44.7	33.5
20	2.0	21.4	28.5	17.1
24	1.7	14.9	19.9	9.9
30	1.3	9.5	12.7	5.1
36	1.1	6.6	9.1[a]	3.0
42	0.9	4.9	7.4[a]	2.1
48	0.8	3.7	6.2[a]	1.6

[a] $L_{ss} = L_{eff} + 2.5$ governs.

6. Compute slenderness ratios, $12L_{ss}/d$. Choices in the range of 3–4 are common.
7. Choose a reasonable size with a diameter and length combination above both the gas capacity and the liquid capacity constraint lines. A 36-in. by 10-ft separator provides about 3 min retention time.

Nomenclature

A_g	cross-sectional area of vessel available for gas settling, ft^2 (m^2)
A_l	cross-sectional area of vessel available for liquid retention, ft^2 (m^2)
A_T	total cross-sectional area of vessel, ft^2 (m^2)
API	API gravity of oil, °API
C_D	drag coefficient, dimensionless
D_m	droplet diameter, ft (m)
D	vessel's internal diameter, ft (m)
d	vessel's internal diameter, in. (mm)
d_m	droplet's diameter, μm
d_{min}	min allowable vessel internal diameter, in. (mm)
d_o	vessel's external diameter, in. (mm)
F_B	buoyant force, lb (N)
F_D	drag force, lb (N)
g	gravitational constant, 32.21 bmft/lbfsec^2 (9.81 m/sec^2)
H	height of liquid volume, ft (m)
h	height of liquid volume, in. (mm)
H_l	height of liquid in horizontal vessel, ft (m)
h_l	height of liquid in horizontal vessel, in. (mm)
L_{eff}	effective length of the vessel, ft (m)
L_{ss}	vessel length seam-to-seam, ft (m)
P	operating pressure, psia (kPa)
Q	flow rate, ft^3/sec (m^3/sec)
Q_g	gas flow rate, MMscfd (std m^3/h)
Q_l	liquid flow rate, BPD (std m^3/h)
Re	Reynolds number, dimensionless
T	operating temperature, °R (K)
t_d	droplet settling time, sec
t_g	gas retention time, sec
t_r	liquid retention time, min
V_g	gas velocity, ft/sec (m/sec)
V_l	average liquid velocity, ft/sec (m/sec)
V_t	terminal settling velocity of the droplet, ft/sec (m/sec)
Z	gas compressibility factor, dimensionless
α	fractional cross-sectional area of liquid, dimensionless

β fractional height of liquid within the vessel $= h_l/d$
ΔSG difference in specific gravity relative to water of the droplet and the gas
$\Delta\rho$ density difference, liquid and gas, lbm/ft^3 (kg/m^3)
μ viscosity, cp
μ_l dynamic viscosity of the liquid, lbm/ftsec (kg/msec)
μ_g gas viscosity, cp (lbsec/ft^2)
ρ density, lb/ft^3 (kg/m^3)
ρ_g density of the gas at the temperature and pressure in the separator, lb/ft^3 (kg/m^3)
ρ_l density of liquid, lb/ft^3 (kg/m^3)

References

Fabian, P., Cusack, R., Hennessey, P., Neuman, M., and van Dessel, P., "Demystifying the Selection of Mist Eliminators," *Chemical Engineering*, Nov. 1993.

Viles, J. C., "Predicting Liquid Re-entrainment in Horizontal Separators" (SPE 25474). Paper presented at the Production Operations Symposium, Oklahoma City, OK, USA, March 1993.

CHAPTER 4

Three-Phase Oil and Water Separators

4.1 Introduction

This chapter discusses the concepts, theory, and sizing equations for the separation of two immiscible liquid phases (in this case, those liquids are normally crude oil and produced water). The separator design concepts presented in Chapter 3 relate to the two-phase separation of liquid and gas and are applicable to the separation of gas that takes place in three-phase separators, gas scrubbers, and any other device in which gas is separated from a liquid phase.

When oil and water are mixed with some intensity and then allowed to settle, a layer of relatively clean free water will appear at the bottom. The growth of this water layer, with time, will follow a curve as shown in Figure 4.1.

After a period of time, ranging anywhere from 3 to 30 min, the change in the water height will be negligible. The water fraction, obtained from gravity settling, is called free water. It is normally beneficial to separate the free water before attempting to treat the remaining oil and emulsion layers.

Three-phase separator and *free-water knockout* are terms used to describe pressure vessels that are designed to separate and remove the free water from a mixture of crude oil and water. Because flow normally enters these vessels directly from either a producing well or a separator operating at a higher pressure, the vessel must be designed to separate the gas that flashes from the liquid, as well as separate the oil and water.

The term *three-phase separator* is normally used when there is a large amount of gas to be separated from the liquid, and the dimensions of the vessel are determined by the gas capacity equations discussed in Chapter 3.

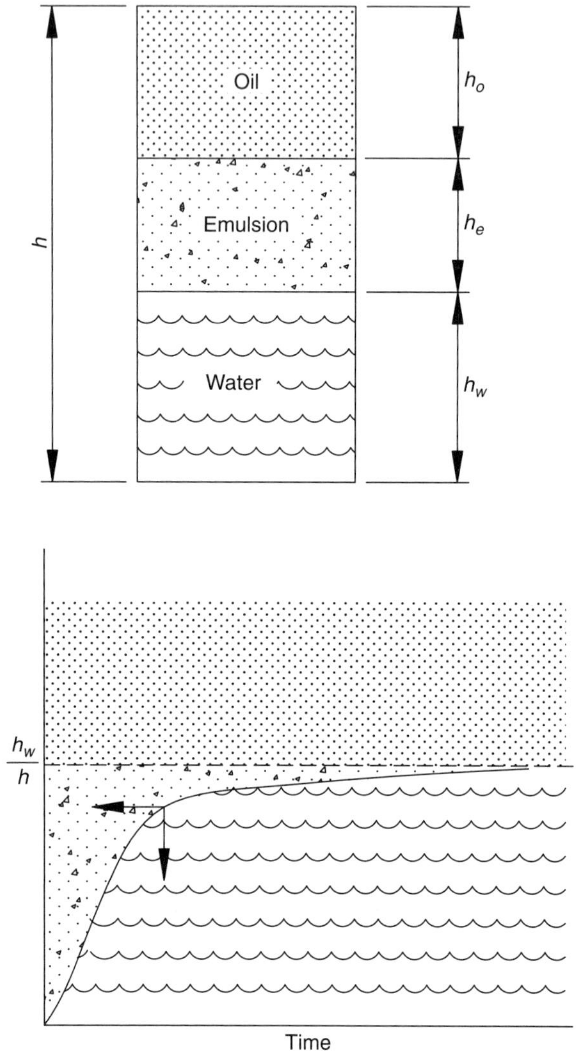

FIGURE 4.1. Growth of water layer with time.

Free-water knockout is generally used when the amount of gas is small relative to the amount of oil and water, and the dimensions of the vessel are determined by the oil–water separation equations discussed in this chapter. No matter what name is given to the vessel, any vessel that is designed to separate two immiscible liquid phases will employ the concepts described in this chapter. For purposes of this chapter, we will call such a vessel a three-phase separator.

The basic design aspects of three-phase separation are identical to those discussed for two-phase separation in Chapter 3. The only additions are that more concern is placed on liquid–liquid settling rates and that some means of removing the free water must be added. Liquid–liquid settling rates will be discussed later in this chapter. Water removal is a function of the control methods used to maintain separation and removal from the oil. Several control methods are applicable to three-phase separators. The shape and diameter of the vessel will, to a degree, determine the types of control used.

4.2 Equipment Description

4.2.1 Horizontal Separators

Three-phase separators are designed as either horizontal or vertical pressure vessels. Figure 4.2 is a schematic of a typical horizontal three-phase separator. The fluid enters the separator and hits an inlet diverter. This sudden change in momentum does the initial gross separation of liquid and vapor as discussed in Chapter 3. In most designs the inlet diverter contains a down-comer that directs the liquid flow below the oil–water interface.

This forces the inlet mixture of oil and water to mix with the water continuous phase in the bottom of the vessel and rise through the oil–water interface. This process is called water washing, and it promotes the coalescence of water droplets, which are entrained in the oil continuous phase.

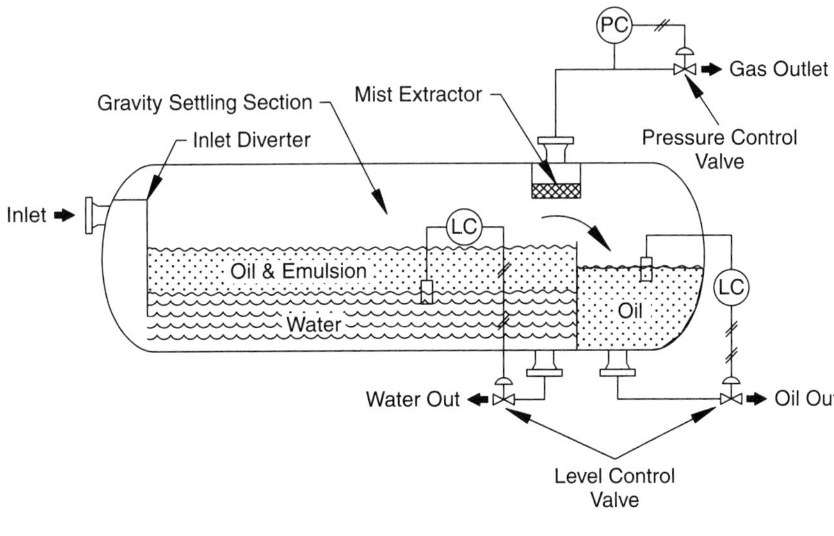

FIGURE 4.2. Schematic of a horizontal three-phase separator with interface level control and weir.

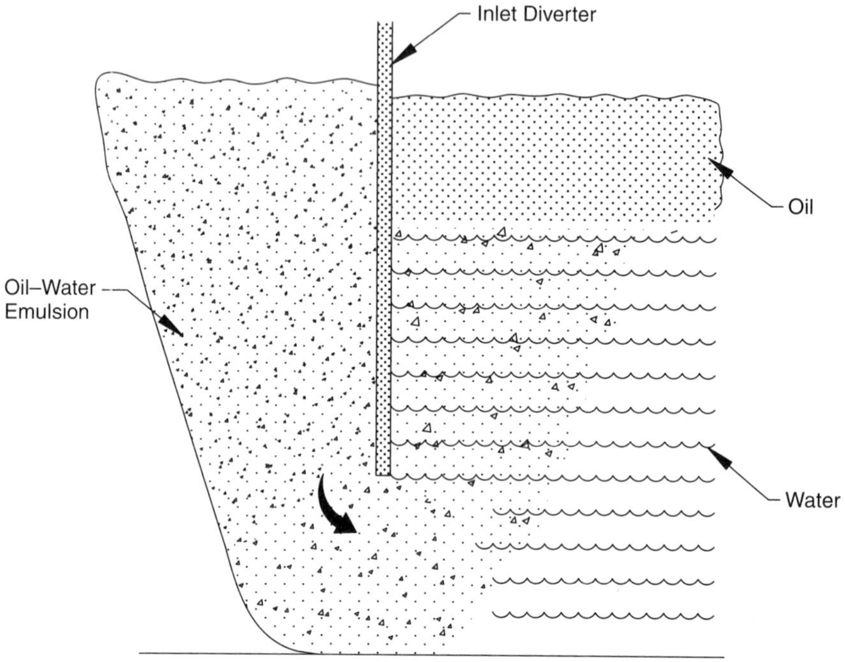

FIGURE 4.3. Inlet diverter illustrating the principles of water washing.

Figure 4.3 illustrates the principles of water washing. The inlet diverter ensures that little gas is carried with the liquid, and the water wash ensures that the liquid does not fall on top of the gas–oil or oil–water interface, mixing the liquid retained in the vessel and making control of the oil–water interface difficult.

The liquid collecting section of the vessel provides sufficient time so that the oil and emulsion form a layer, or oil pad, on top of the free water. The free water settles to the bottom.

Figure 4.4 is a cutaway view of a typical horizontal three-phase separator with an interface level controller and weir. The weir maintains the oil level, and the level controller maintains the water level. The oil is skimmed over the weir. The level of the oil downstream of the weir is controlled by a level controller that operates the oil dump valve.

The produced water flows from a nozzle in the vessel located upstream of the oil weir. An interface level controller senses the height of the oil–water interface. The controller sends a signal to the water dump valve, thus allowing the correct amount of water to leave the vessel so that the oil–water interface is maintained at the design height.

The gas flows horizontally and out through a mist extractor to a pressure control valve that maintains constant vessel pressure. The

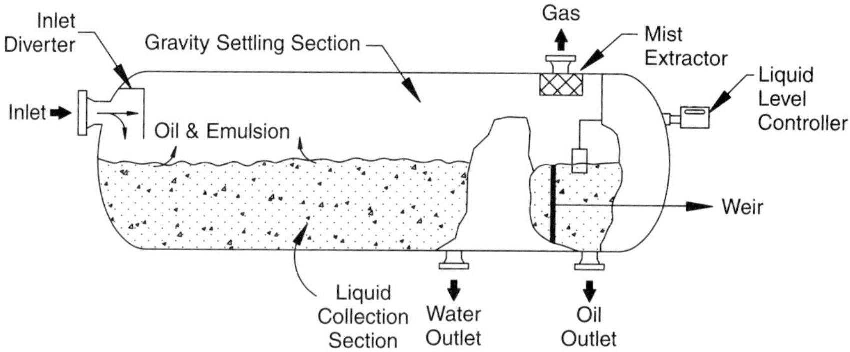

FIGURE 4.4. Cutaway view of a horizontal three-phase separator with interface level control and weir.

level of the gas–oil interface can vary from 50% to 75% of the diameter depending on the relative importance of liquid–gas separation. The most common configuration is half-full, and this is used for the design equations in this section. Similar equations can be developed for other interface levels.

Figure 4.5 shows an alternate configuration known as a "bucket and weir" design. Figure 4.6 is a cutaway view of a horizontal three-phase separator with a bucket and weir. This design eliminates the need for a liquid interface controller. Both the oil and water flow over

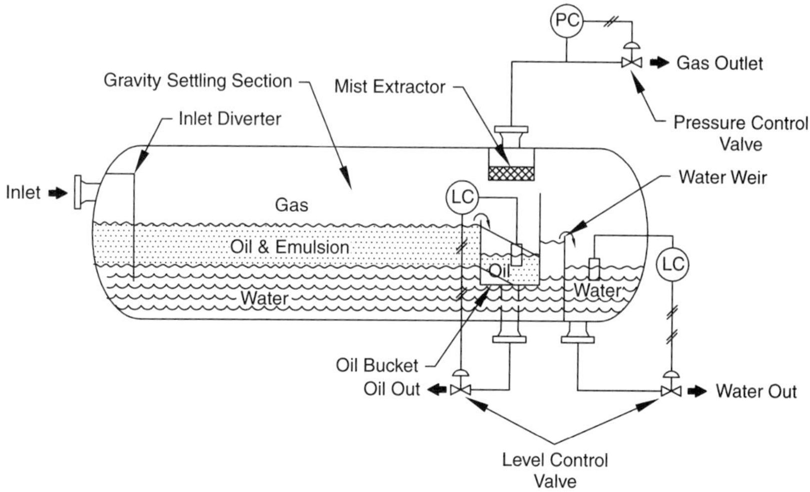

FIGURE 4.5. Schematic of a horizontal three-phase separator with a bucket and weir.

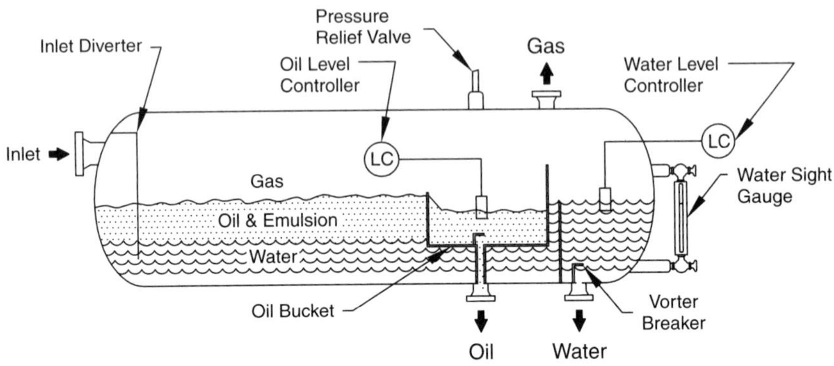

FIGURE 4.6. Cutaway view of a horizontal three-phase separator with a bucket and weir.

weirs where level control is accomplished by a simple displacer float. The oil overflows the oil weir into an oil bucket where its level is controlled by a level controller that operates the oil dump valve. The water flows under the oil bucket and then over a water weir. The level downstream of this weir is controlled by a level controller that operates the water dump valve.

As shown in Figures 4.5 and 4.6, the back of the oil bucket is higher than the front of the bucket. This differential height configuration assures oil will not flow over the back of the bucket and out with the water should the bucket become flooded (Figure 4.7). The height of the oil weir controls the liquid level in the vessel. The difference in height of the oil and water weirs controls the thickness of the oil pad due to specific gravity differences. It is critical to the operation

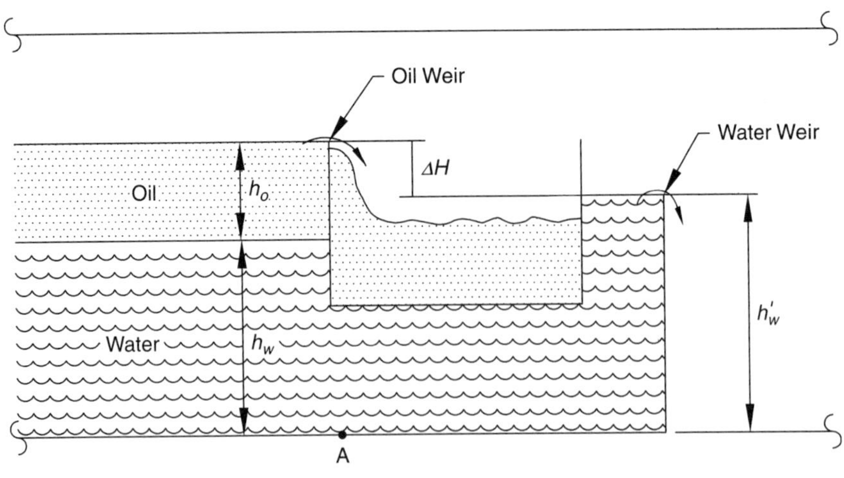

FIGURE 4.7. Determination of oil pad height.

of the vessel that the water weir height is sufficiently below the oil weir height so that the oil pad thickness provides sufficient oil retention time. If the water weir is too low and the difference in specific gravity is not as great as anticipated, then the oil pad could grow in thickness to a point where oil will be swept under the oil box and out the water outlet. Normally, either the oil or the water weir is made adjustable so that changes in oil- or water-specific gravities or flow rates can be accommodated.

To obtain a desired oil pad height, the water weir should be set a distance below the oil weir. This distance is calculated by using Equation (4.1), which is developed by equating the static heads at point A.

$$\Delta h = h_o \left[1 - \left(\frac{\rho_o}{\rho_w} \right) \right] \tag{4.1}$$

where Δh = distance below the oil weir, in (mm), h_o = desired oil pad height, in (mm), ρ_o = oil density, lb/ft^3 (kg/m^3), ρ_w = water density, lb/ft^3 (kg/m^3).

This equation neglects the height of the oil and water flowing over the weir and presents a view of the levels when there is no inflow. A large inflow of oil will cause the top of the oil pad to rise; the oil pad will thus get thicker, and the oil bucket must be deep enough so that oil does not flow under it. Similarly, a large inflow of water will cause the level of water flowing over the water weir to rise, and there will be a large flow of oil from the oil pad over the oil weir until a new h_w is established. These dynamic effects can be minimized by making the weirs as long as possible.

Three-phase separators with a bucket and weir design are most effective with high water-to-oil flow rates and/or small density differences. Interface control design has the advantage of being easily adjustable to handle unexpected changes in oil or water specific gravity or flow rates. Interface control should be considered for applications with high oil flow rates and/or large density differences. However, in heavy oil applications or where large amounts of emulsion or paraffin are anticipated, it may be difficult to sense interface level. In such a case bucket and weir control is recommended.

Free-Water Knockout

The term *free-water knockout* (FWKO) is reserved for a vessel that processes an inlet liquid stream with little entrained gas and makes no attempt to separate the gas from the oil. Figure 4.8 illustrates a horizontal FWKO.

Figure 4.9 illustrates a vertical FWKO. The major difference between a conventional three-phase separator and an FWKO is that in the latter there are only two fluid outlets; one for oil and very small amounts of gas and the second for the water. FWKOs are usually operated as packed vessels. Water outflow is usually controlled with an interface

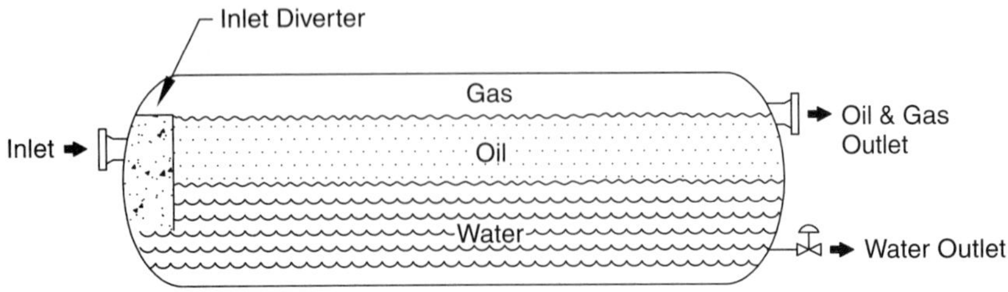

FIGURE 4.8. Schematic of a horizontal FWKO.

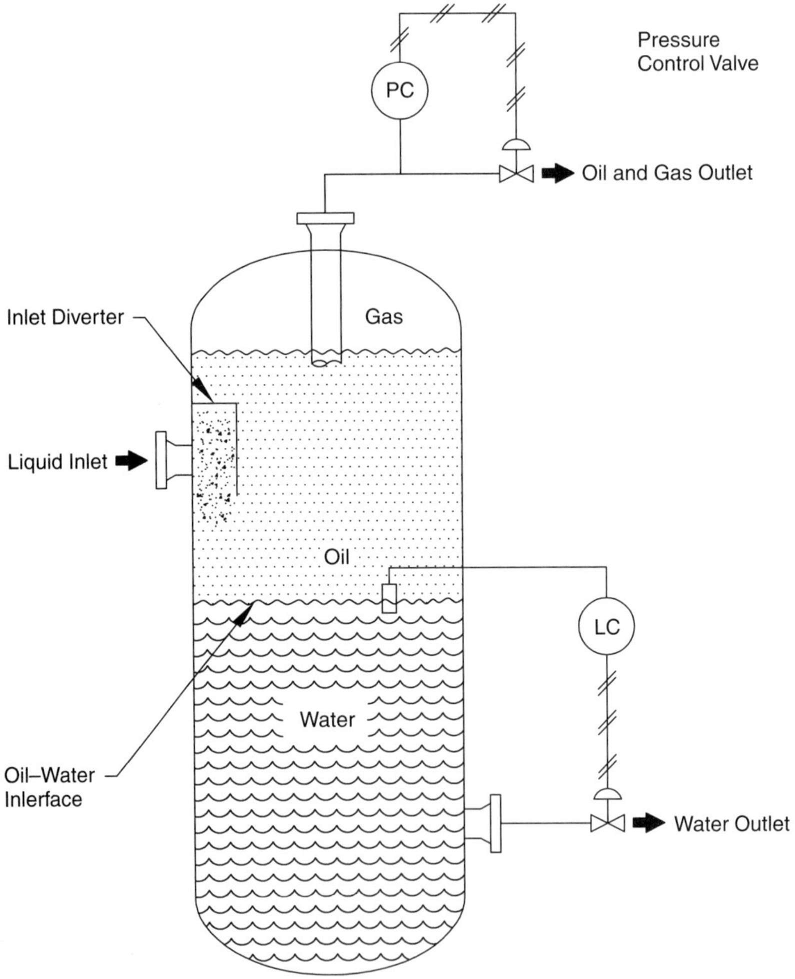

FIGURE 4.9. Schematic of a vertical FWKO.

level control. It should be clear that the principles of operation of such a vessel are the same as those described above. The design of an FWKO is the same as that of a three-phase separator. Since there is very little gas, the liquid capacity constraint always dictates the size.

Flow Splitter

Figure 4.10 illustrates a typical flow splitter. A flow splitter is a special version of a free-water knockout. Basically, it is an FWKO where the oil outlet is split among two or more outlet lines that are directed to several downstream process components. This vessel contains several compartments, which are sealed from each other. Each compartment has its own level control and outlet oil valve. Unlike the FWKO, which may be operated as a packed vessel, the flow splitter must be operated with a gas blanket. Adjustable weirs separate the compartments from water and oil outside the compartments.

Oil flows over the weirs into the individual compartments. The water level control is used to maintain the top of the oil layer above the highest weir. Individual level controls in each compartment ensure that the oil leaves the compartments at the same rate at which it enters. The flow of liquid across the notched weir is directly proportional to the difference in height between the liquid upstream of the weir and the bottom of the notch. When the weirs of different compartments are set at different heights, the flow into each compartment is different. The water level control holds the water level constant, which ensures all oil that enters the separator leaves through the compartments in proportions related to the weir heights.

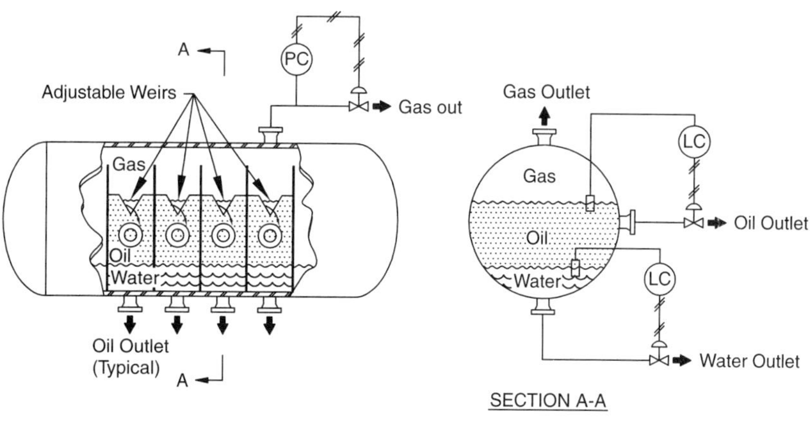

FIGURE 4.10. Schematic of a flow splitter with four compartments.

Horizontal Three-Phase Separator with a Liquid Boot

Figure 4.11 shows a horizontal three-phase separator with a water "boot" on the bottom of the vessel barrel. The boot collects small amounts of water that settle out in the liquid collection section and travel to the outlet end of the vessel. These vessels are a special case of three-phase separators.

Figure 4.12 shows a horizontal two-phase separator with a liquid boot. Because the water flow rate is so low relative to the oil flow rate,

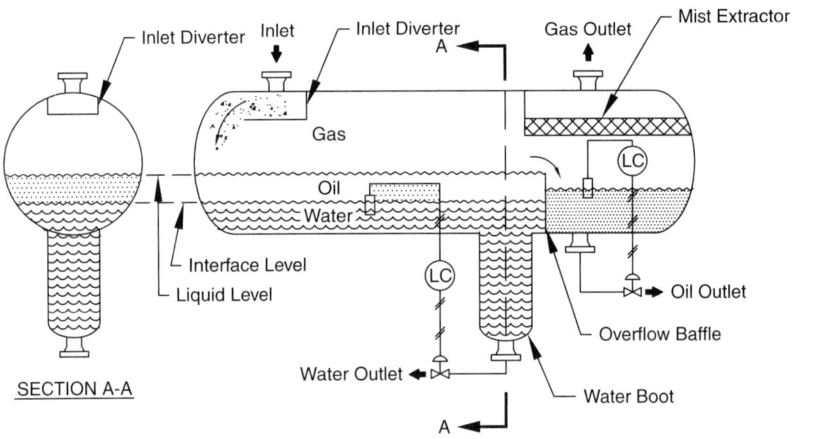

FIGURE 4.11. Schematic of a horizontal three-phase separator with a water boot.

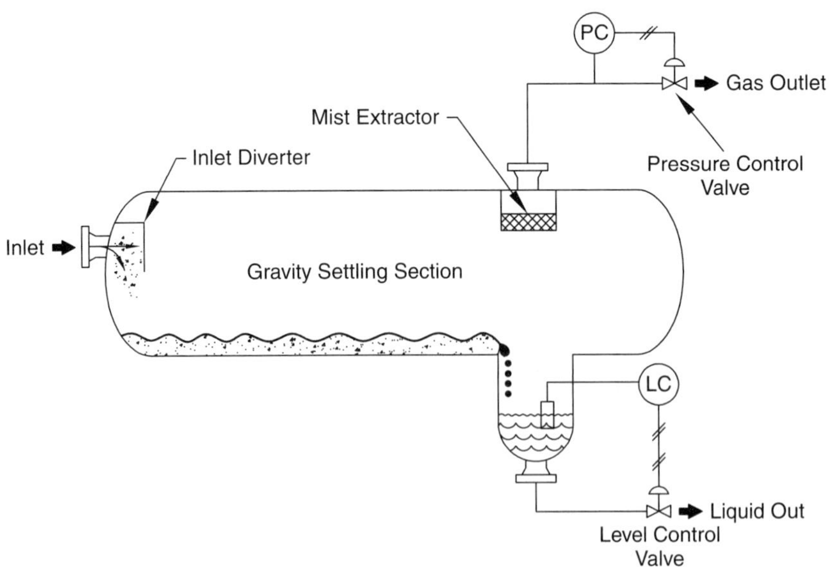

FIGURE 4.12. Schematic of a horizontal two-phase separator with a liquid boot.

the small amount of water retention time provided by the boot is sufficient. Thus the diameter of the main body of the vessel can be smaller. The liquid boot collects small amounts of liquid in the liquid collection section. These vessels are a special case of two-barrel two--phase separators, which are typically used in dry gas applications and should only be used where separation of the two liquid phases is relatively easy.

4.3 Vertical Separators

Figure 4.13 shows a typical configuration for a vertical three-phase separator. Flow enters the vessel through the side as in the horizontal separator. The inlet diverter separates the bulk of the gas. A down-comer is required to route the liquid through the oil–gas interface so as not to disturb the oil skimming action taking place. A chimney is

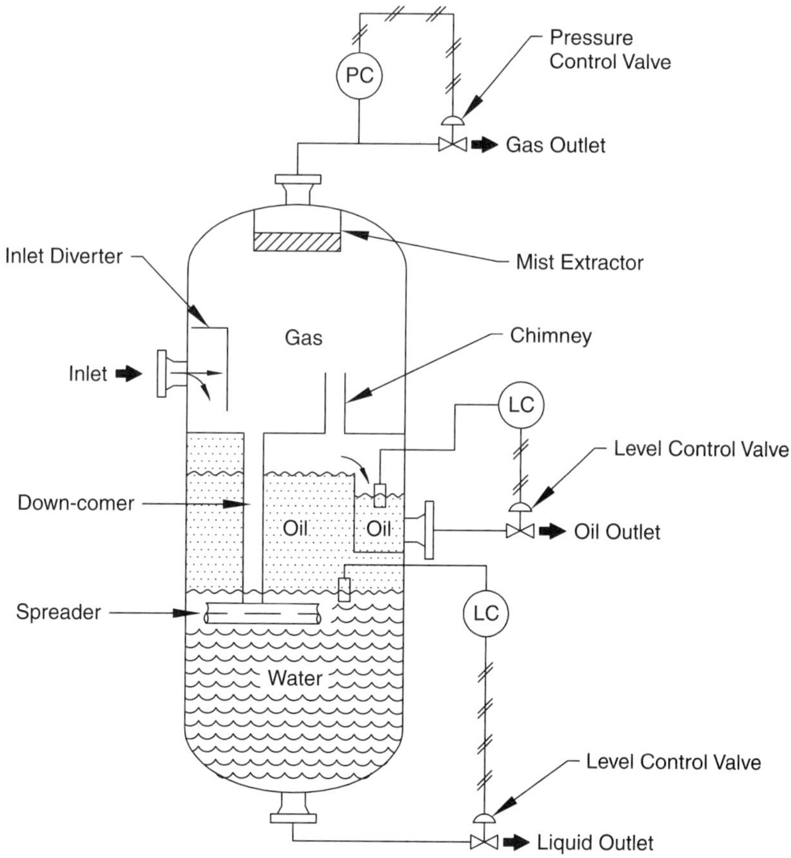

FIGURE 4.13. Schematic of a vertical three-phase separator with interface level control.

needed to equalize gas pressure between the lower section and the gas section.

The spreader, or down-comer, outlet is located just below the oil–water interface, thus water washing the incoming stream. From this point, as the oil rises, any free water trapped within the oil phase separates out. The water droplets flow countercurrent to the oil. Similarly, the water flows downward, and oil droplets trapped in the water phase tend to rise countercurrent to the water flow. Figures 4.14 and 4.15 are views of vertical three-phase separators without water washing and with interface control.

Figure 4.16 shows the three different methods of control that are often used on vertical separators.

- The first is strictly level control. A regular displacer float is used to control the gas–oil interface and regulate a control valve dumping oil from the oil section. An interface float is used to control the oil–water interface and regulate a water outlet control valve. Because no internal baffling or weirs are

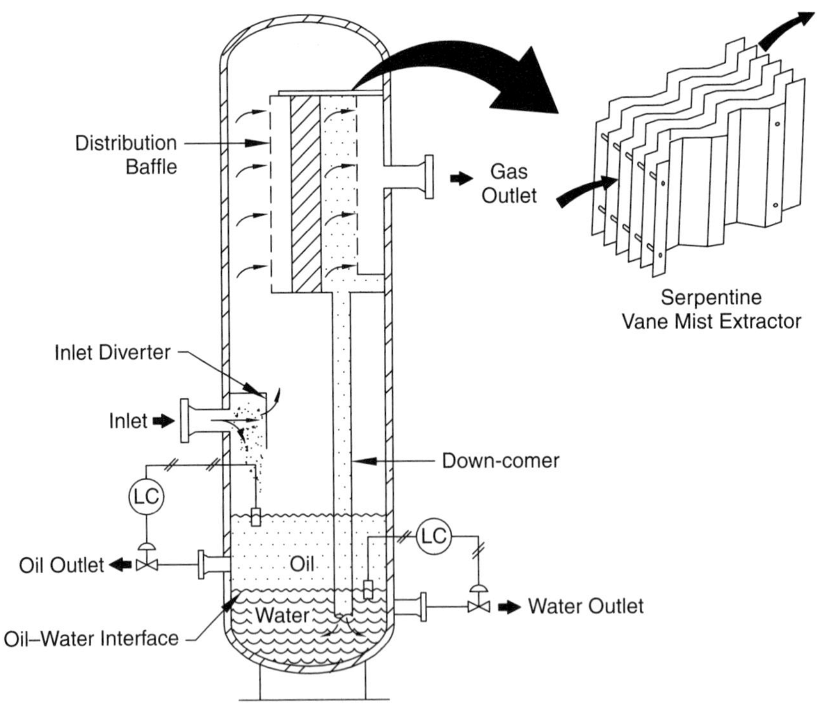

FIGURE 4.14. Cutaway view of a vertical three-phase separator with interface level control.

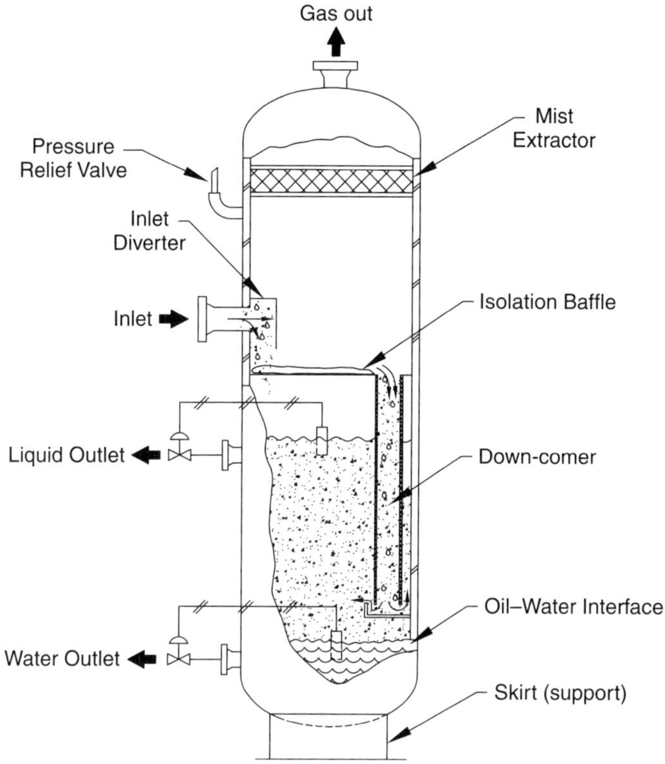

FIGURE 4.15. Cutaway view of a vertical three-phase separator without water washing.

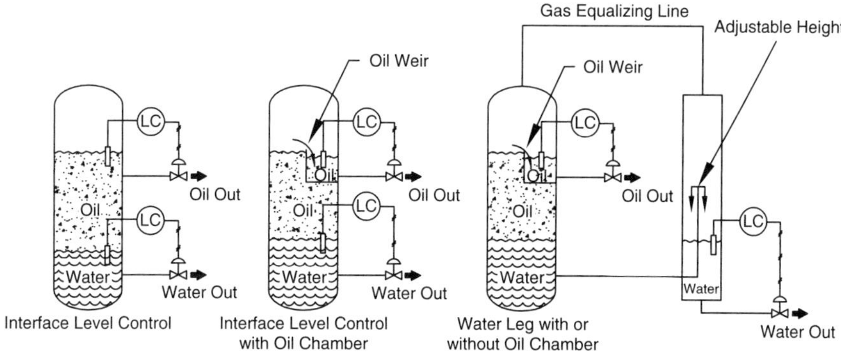

FIGURE 4.16. Liquid level control schemes.

used, this system is the easiest to fabricate and handles sand and solids production best.

- The second method shown uses a weir to control the gas–oil interface level at a constant position. This results in a better separation of water from the oil as all the oil must rise to the height of the oil weir before exiting the vessel. Its disadvantages are that the oil box takes up vessel volume and costs money to fabricate. In addition, sediment and solids could collect in the oil box and be difficult to drain, and a separate low-level shut-down may be required to guard against the oil dump valve's failing to close.
- The third method uses two weirs, which eliminates the need for an interface float. Interface level is controlled by the height of the external water weir relative to the oil weir or outlet height. This is similar to the bucket and weir design of horizontal separators. The advantage of this system is that it eliminates the interface level control. The disadvantage is that it requires additional external piping and space. In cold climates the water leg is sometimes installed internal to the vessel so that the vessel insulation will prevent it from freezing.

4.4 Selection Considerations

The geometry and physical and operating characteristics give each separator type advantages and disadvantages. Gravity separation is more efficient in horizontal vessels than in vertical vessels. In the gravity settling section of a horizontal vessel, the settling velocity and flow velocity are perpendicular rather than countercurrent in a vertical vessel.

Horizontal separators have greater interface areas, which enhances phase equilibrium. This is especially true if foam or emulsion collect at the gas–oil interface. Thus, from a process perspective, horizontal vessels are preferred. However, they do have several drawbacks, which could lead to a preference for a vertical vessel in certain situations:

1. Horizontal separators are not as good as vertical separators in handling solids. The liquid dump valve of a vertical separator can be placed at the center of the bottom head so that solids will not build up in the separator, but continue to the next vessel in the process. As an alternative, a drain could be placed at this location so that solids could be disposed of periodically while liquid leaves the vessel at a slightly higher elevation. In a horizontal vessel, it is necessary to place several drains along the length of the vessel. Since the solids will have an angle of repose

of 45° to 60°, the drains must be spaced at very close intervals [usually no farther than 5 ft (1.5 m) apart]. Attempts to lengthen the distance between drains, by providing sand jets in the vicinity of each drain to fluidize the solids while the drains are in the operation, are expensive and have been only marginally successful in field operations.

2. Horizontal vessels require more plan area to perform the same separation as vertical vessels. While this may not be of importance at a land location, it could be very important offshore. If several separators are used, however, this disadvantage may be overcome by stacking horizontal separators on top of each other.
3. Small-diameter horizontal vessels [3-ft (1.5-m) diameter and smaller] have less liquid surge capacity than vertical vessels sized for the same steady-state flow rate. For a given change in liquid surface elevation, there is typically a larger increase in liquid volume for a horizontal separator than for a vertical separator sized for the same flow rate. However, the geometry of a small horizontal vessel causes any high-level shutdown device to be located close to the normal operating level. In very large diameter [greater than 6 ft (1.8 m)] horizontal vessels and in vertical vessels, the shutdown could be placed much higher, allowing the level controller and dump valve more time to react to the surge. In addition, surges in horizontal vessels could create internal waves, which could activate a high-level sensor prematurely.
4. Care should be exercised when selecting small-diameter [5 ft (1.5 m)] horizontal separators. The level controller and level switch elevations must be considered. The vessel must have a sufficiently large diameter so that the level switches may be spaced far enough apart, vertically, so as to avoid operating problems. This is important if surges in the flow of slugs of liquids are expected to enter the separator.

It should be pointed out that vertical vessels have some drawbacks that are not process related and that must be considered when making a selection. For example, the relief valve and some of the controls may be difficult to service without special ladders and platforms. The vessel may have to be removed from the skid for trucking due to height restrictions.

In summary, horizontal vessels are most economical for normal oil–water separation, particularly where there may be problems with emulsions, foam, or high gas–liquid ratios. Vertical vessels work most effectively in low gas–oil ratio (GOR) applications and where solids production is anticipated.

4.5 Vessel Internals

Vessel internals common to both two-phase and three-phase separators, such as inlet diverters, wave breakers, defoaming plates, vortex breakers, stilling wells, sand jets and drains, and mist extractors, are covered in Chapter 3: Two-Phase Oil and Gas Separation and will not be repeated here. Additional internals that aid in the separation of oil and water are presented in this section.

4.5.1 Coalescing Plates

It is possible to use various plate or pipe coalescer designs to aid in the coalescing of oil droplets in the water and water droplets in the oil. The installation of coalescing plates in the liquid section will cause the size of the water droplets entrained in the oil phase to increase, making gravity settling of these drops to the oil–water interface easier. Thus, the use of coalescing plates (Figure 4.17), will often lead to the ability to handle a given flow rate in a smaller vessel. However, because of the potential for plugging with sand, paraffin, or corrosion products, the use of coalescing plates should be discouraged, except for instances where the savings in vessel size and weight are large enough to justify the potential increase in operating costs and decrease in availability.

4.5.2 Turbulent Flow Coalescers

Turbulent flow coalescers, which were marketed under the name SP Packs, utilized the turbulence created by flow in a serpentine pipe path to promote coalescence.

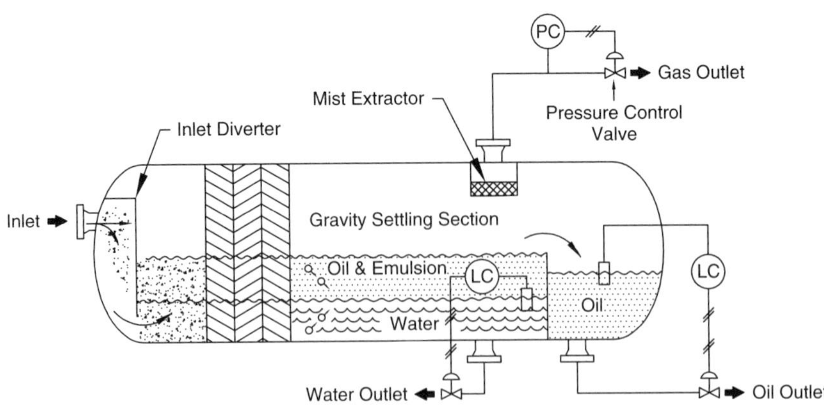

FIGURE 4.17. Schematic of a horizontal three-phase separator fitted with coalescing plates.

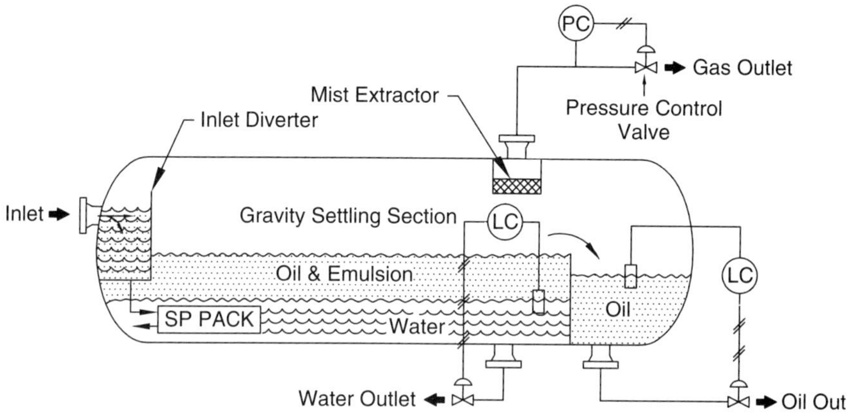

FIGURE 4.18. Schematic of a horizontal three-phase separator fitted with a free-flow turbulent coalescers (SP Packs).

As shown in Figure 4.18, SP Packs took up more space in the vessel than plate coalescers, but since they did not have small clearances, they were not susceptible to plugging. Despite the design advantages, the units were not well received and, as such, are no longer being manufactured.

4.6 Potential Operating Problems

Emulsions. Three-phase separators may experience the same operating problems as two-phase separators. In addition, three-phase separators may develop problems with emulsions which can be particularly troublesome in the operation of three-phase separators. Over a period of time an accumulation of emulsified materials and/or other impurities may form at the interface of the water and oil phases. In addition to adverse effects on the liquid level control, this accumulation will also decrease the effective oil or water retention time in the separator, with a resultant decrease in water–oil separation efficiency. Addition of chemicals and/or heat often minimizes this difficulty.

Frequently, it is possible to appreciably lower the settling time necessary for water–oil separation by either the application of heat in the liquid section of the separator or the addition of de-emulsifying chemicals.

4.7 Design Theory

Gas separation. The concepts and equations pertaining to two-phase separation described in Chapter 3 are equally valid for three-phase separation.

4.7.1 Oil–Water Settling

It can be shown that flow around settling oil drops in water or water drops in oil is laminar and thus Stokes' law governs. The terminal drop velocity is

Field units

$$V_t = \frac{1.78 \times 10^{-6}(\Delta SG)d_m^2}{\mu} \tag{4.2a}$$

SI units

$$V_t = \frac{5.56 \times 10^{-7}(\Delta SG)d_m^2}{\mu} \tag{4.2b}$$

where V_t = terminal settling velocity, ft/s (m/s), ΔSG = difference in specific gravity relative to water between the oil and the water phases, d_m = drop size, μm, μ = viscosity of continuous phase, cp.

4.7.2 Water Droplet Size in Oil

It is difficult to predict the water droplet size that must be settled out of the oil phase to coincide with the rather loose definition of "free oil." Unless laboratory or nearby field data are available, good results have been obtained by sizing the oil pad such that water droplets 500 μm and larger settle out.

As shown in Figure 4.19, if this criterion is met, the emulsion to be treated by downstream equipment should contain less than 5–10% water. In heavy crude oil systems, it is sometimes necessary to design for 1000-μm water droplets to settle. In such cases the emulsion may contain as much as 20–30% water.

4.7.3 Oil Droplet Size in Water

From Equations (4.2a) and (4.2b) it can be seen that the separation of oil droplets from the water is easier than the separation of water droplets from the oil. The oil's viscosity is on the order of 5–20 times that of water.

Thus, the terminal settling velocity of an oil droplet in water is much larger than that of a water droplet in oil. The primary purpose of three-phase separation is to prepare the oil for further treating. Field experience indicates that oil content in the produced water from a three-phase separator, sized for water removal from oil, can be expected to be between a few hundred and 2000 mg/l. This water will require further treating prior to disposal. Sizing for oil droplet

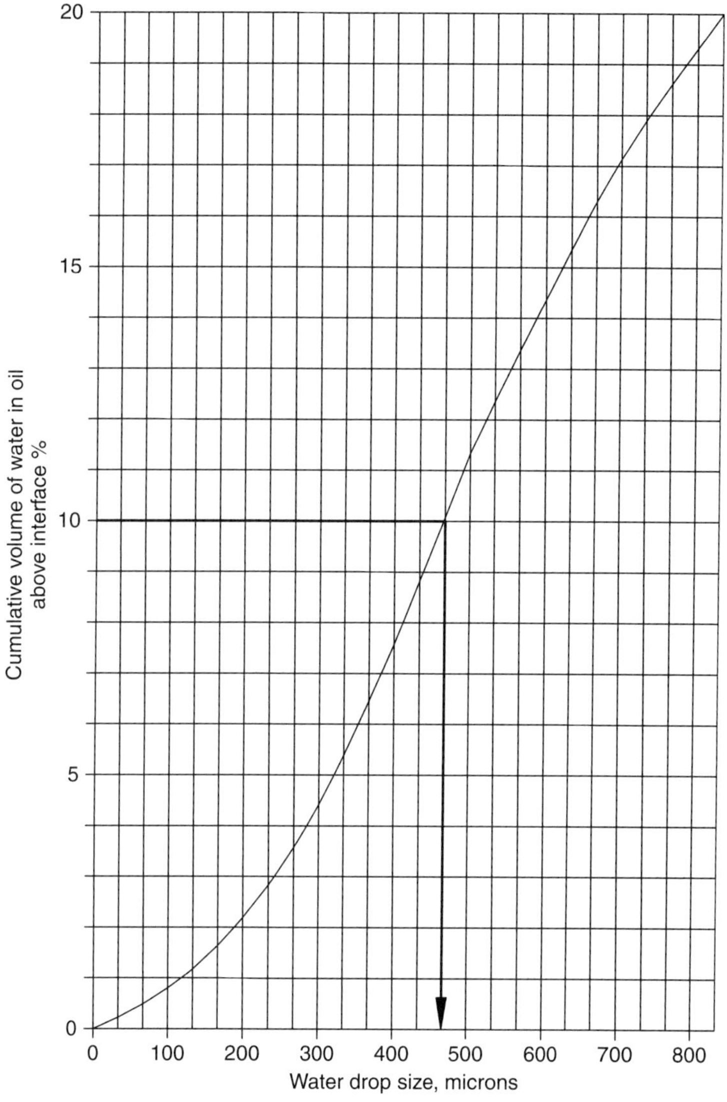

FIGURE 4.19. Typical water droplet size distribution.

removal from the water phase does not appear to be a meaningful criterion.

Occasionally, the viscosity of the water phase may be as high as, or higher than, the liquid hydrocarbon phase viscosity. For example, large glycol dehydration systems usually have a three-phase flash separator. The viscosity of the glycol/water phase may be rather high. In cases like this, the settling equation should be applied to removing oil droplets of approximately 200 μm from the water phase.

If the retention time of the water phase is significantly less than the oil phase, then the vessel size should be checked for oil removal from the water. For these reasons, the equations are provided so the water phase may be checked. However, the separation of oil from the water phase rarely governs the vessel size and may be ignored for most cases.

4.7.4 Retention Time

A certain amount of oil storage is required to ensure that the oil reaches equilibrium and that flashed gas is liberated. An additional amount of storage is required to ensure that the free water has time to coalesce into droplet sizes sufficient to fall in accordance with Equations (4.2a) and (4.2b). It is common to use retention times ranging from 3 to 30 min depending on laboratory or field data. If this information is not available, the guidelines presented in Table 4.1 can be used.

Generally, the retention time must be increased as the oil gravity or viscosity increases. Similarly, a certain amount of water storage is required to ensure that most of the large droplets of oil entrained in the water have sufficient time to coalesce and rise to the oil–water interface. It is common to use retention times for the water phase ranging from 3 to 30 min depending on laboratory or field data. If this information is not available, a water retention time of 10 min is recommended for design.

The retention time for both the maximum oil rate and the maximum water rate should be calculated, unless laboratory data indicate that it is unnecessary to take this conservative design approach.

TABLE 4.1
Oil retention time

°API Gravity	*Time (Min)*
Condensate	2–5
Light crude oil (30°–40°)	5–7.5
Intermediate crude oil (20°–30°)	7.5–10
Heavy crude oil (less than 20°)	10+

Note: If an emulsion exists in inlet stream, increase above retention times by a factor of 2–4.

4.8 Separator Design

The guidelines presented here can be used for initial sizing of a horizontal three-phase separator 50% full of liquid. They are meant to complement, and not replace, operating experiences. Determination of the type and size of the separator must be made on an individual basis. All the functions and requirements should be considered including the likely uncertainties in design flow rates and properties. For this reason, there is no substitute for good engineering evaluations of each separator by the design engineer. The trade-off between design size and details and uncertainties in design parameters should not be left to manufacturer recommendations or rules of thumb.

4.8.1 Horizontal Separator Sizing—Half-Full

For sizing a horizontal three-phase separator it is necessary to specify a vessel diameter and a seam-to-seam vessel length. The gas capacity and retention time considerations establish certain acceptable combinations of diameter and length. The need to settle 500-μm water droplets from the oil and 200-μm oil droplets from the water establishes a maximum diameter corresponding to the given liquid retention time.

Gas Capacity Constraint

The principles of liquid droplets settling through a gas were given in Chapter 3. By setting the gas retention time equal to the time required for a drop to settle to the liquid interface, the following equations may be derived:

Field units

$$dL_{\text{eff}} = 420\left[\frac{TZQ_g}{P}\right]\left[\left(\frac{\rho_g}{\rho_l - \rho_g}\right)\frac{C_D}{d_m}\right]^{1/2} \tag{4.3a}$$

SI units

$$dL_{\text{eff}} = 34.5\left[\frac{TZQ_g}{P}\right]\left[\left(\frac{\rho_g}{\rho_l - \rho_g}\right)\frac{C_D}{d_m}\right]^{1/2} \tag{4.3b}$$

where d = vessel inside diameter, in. (mm), L_{eff} = vessel effective length, ft (m), T = operating temperature, °R (°K), Z = gas compressibility, Q_g = gas flow rate, MMscfd (scm/h), P = operating pressure, psia (kPa), p_g = density of gas, lb/ft^3 (kg/m^3), p_l = density of liquid, lb/ft^3 (kg/m^3), C_D = drag coefficient, d_m = liquid drop to be separated, μm.

Retention Time Constraint

Liquid retention time constraints can be used to develop the following equation, which may be used to determine acceptable combinations of d and L_{eff}

Field units

$$d^2 L_{eff} = 1.42[(Q_W)(t_r)_w + (Q_O)(t_r)_O] \quad (4.4a)$$

SI units

$$d^2 L_{eff} = 4.2 \times 10^4 [(Q_W)(t_r)_w + (Q_O)(t_r)_O] \quad (4.4b)$$

where Q_W = water flow rate, BPD (m^3/h), $(t_r)_W$ = water retention time, min, Q_o = oil flow rate, BPD (m^3/h), $(t_r)_o$ = oil retention time, min.

Settling Water Droplets from Oil Phase

The velocity of water droplets settling through oil can be calculated using Stokes' law. From this velocity and the specified oil phase retention time, the distance that a water droplet can settle may be determined. This settling distance establishes a maximum oil pad thickness given by the following formula:

Field units

$$h_o = \frac{0.00128(t_r)_o(\Delta SG)d_m^2}{\mu} \quad (4.5a)$$

SI units

$$h_o = \frac{0.0033(t_r)_o(\Delta SG)d_m^2}{\mu} \quad (4.5b)$$

This is the maximum thickness the oil pad can be and still allow the water droplets to settle out in time $(t_r)_o$. For $d_m = 500\ \mu m$, the following equation may be used.

Field units

$$(h_o)_{max} = 320\frac{(t_r)_o(\Delta SG)}{\mu} \quad (4.6a)$$

SI units

$$(h_o)_{max} = 8250\frac{(t_r)_o(\Delta SG)}{\mu} \quad (4.6b)$$

For a given oil retention time $[(t_r)_o]$ and a given water retention time $[(t_r)_W]$, the maximum oil pad thickness constraint establishes a maximum diameter in accordance with the following procedure:

1. Compute (h_o)max- Use 500-μm droplet if no other information is available.
2. Calculate the fraction of the vessel cross-sectional area occupied by the water phase. This is given by

$$\frac{A_w}{A} = 0.5 \frac{Q_w (t_r)_w}{(t_r)_o Q_o + (t_r)_w Q_w} \tag{4.7}$$

3. From Figure 4.20, determine the coefficient β.
4. Calculate d_{max} from

$$d_{max} = \frac{(h_o)max}{\beta} \tag{4.8}$$

Any combination of d and L_{eff} that satisfies all three of Equations (4.3), (4.4), and (4.8) will meet the necessary criteria.

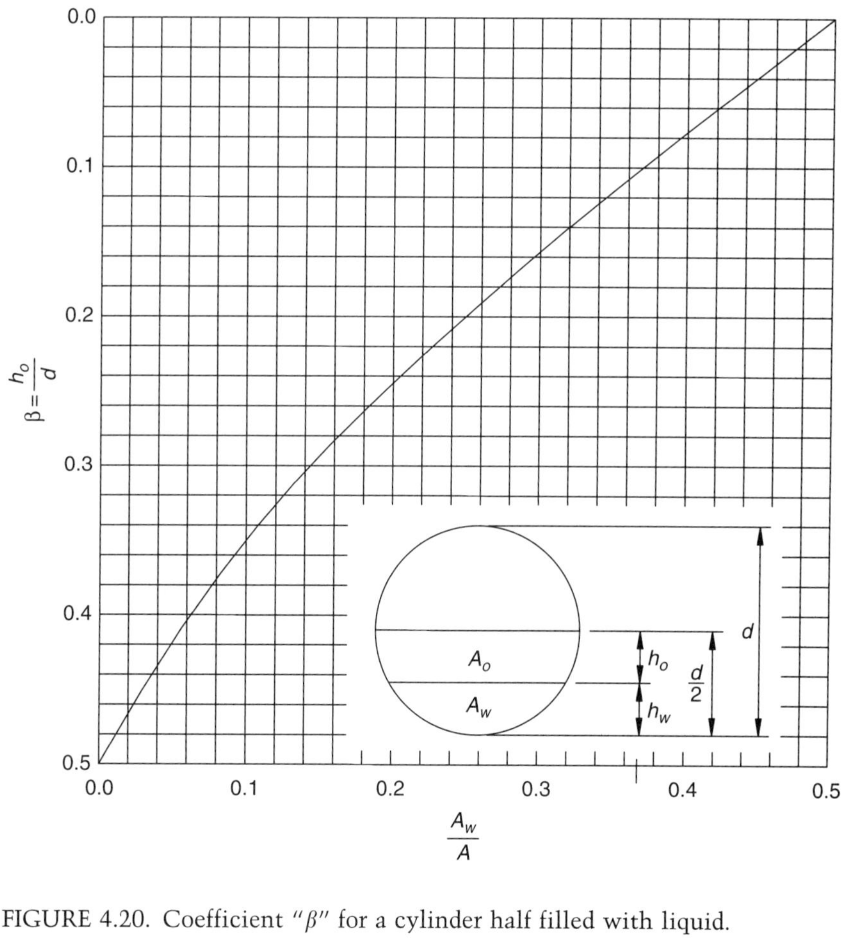

FIGURE 4.20. Coefficient "β" for a cylinder half filled with liquid.

4.9 Separating Oil Droplets from Water Phase

Oil droplets in the water phase rise at a terminal velocity defined by Stokes' law. As with water droplets in oil, the velocity and retention time may be used to determine a maximum vessel diameter. It is rare that the maximum diameter determined from a 200-μm oil droplet rising through the water phase is larger than a 500-μm water droplet falling through the oil phase. Therefore, the maximum diameter determined from a 500-μm water droplet settling through the oil phase normally governs the vessel design. For $d_m = 200$ μm, the following equations may be used:

Field units

$$(h_w)_{max} = \frac{(51.2(t_r)_w(\Delta SG))}{\mu_w} \tag{4.9a}$$

SI units

$$(h_w)_{max} = \frac{(1,520(t_r)_w(\Delta SG))}{\mu_w} \tag{4.9b}$$

The maximum diameter may be found from the following equation:

$$d_{max} = \frac{(h_w)_{max}}{\beta} \tag{4.10}$$

4.9.1 Seam-to-Seam Length

The effective length may be calculated from Equations (4.4a) and (4.4b). From this, a vessel seam-to-seam length may be estimated. The actual required seam-to-seam length is dependent on the physical design of the vessel.

For vessels sized based on gas capacity, some portion of the vessel length is required to distribute the flow evenly near the inlet diverter. Another portion of the vessel length is required for the mist extractor. The length of the vessel between the inlet and the mist extractor with evenly distributed flow is the L_{eff} calculated from Equations (4.3a) and (4.3b).

As a vessel's diameter increases, more length is required to evenly distribute the gas flow. However, no matter how small the diameter may be, a portion of the length is still required for the mist extractor and flow distribution. Based on these concepts coupled with field experience, the seam-to-seam length of a vessel may be estimated as the larger of the following:

$$L_{ss} = \frac{4}{3}L_{eff} \tag{4.11}$$

Field units

$$L_{ss} = L_{eff} + d/12 \quad (4.12a)$$

SI units

$$L_{ss} = L_{eff} + d/1000 \quad (4.12b)$$

For vessels sized on a liquid capacity basis, some portion of the vessel length is required for inlet diverter flow distribution and liquid outlet. The seam-to-seam length should not exceed the following:

$$L_{ss} = 4/3L_{eff} \quad (4.13)$$

4.9.2 Slenderness Ratio

For each vessel design, a combination of L_{eff} and d exists that will minimize the cost of the vessel. In general, the smaller the diameter of a vessel, the less it will cost. However, decreasing the diameter increases the fluid velocities and turbulence. As a vessel diameter decreases, the likelihood of the gas re-entraining liquids or destruction of the oil/water interface increases. Experience indicates that the ratio of the seam-to-seam length divided by the outside diameter should be between 3 and 5. This ratio is referred to as the 'slenderness ratio' (SR) of the vessel. Slenderness ratios outside the 3–5 range may be used but are not as common. Slenderness ratios outside the 3–5 range may be used, but the design should be checked to assure that re-entrainment will not occur.

4.9.3 Procedure for Sizing Three-Phase Horizontal Separators—Half-Full

1. The first step in sizing a horizontal separator is to establish the design basis. This includes specifying the maximum and minimum flow rates, operating pressure and temperature, droplet size to be removed, and so on.
2. Select a $(t_r)_o$ and a $(t_r)_w$.
3. Calculate $(h_o)_{max}$. Use a 500-μm droplet if no other information is available.

Field units

$$(h_o)_{max} = 1.28 \times 10^{-3} \frac{(t_r)_o(\Delta SG)d_m^2}{\mu}$$

For 500 μm,

$$(h_o)_{max} = 320 \frac{(t_r)_o(\Delta SG)}{\mu}$$

SI units

$$(h_o)_{max} = 0.033\frac{(t_r)_o(\Delta SG)d_m^2}{\mu}$$

For 500 μm,

$$(h_o)_{max} = 8250\frac{(t_r)_o(\Delta SG)}{\mu}$$

4. Calculate A_w/A:

$$\frac{A_w}{A} = 0.5\frac{Q_w(t_r)_w}{(t_r)_o Q_o + (t_r)_w Q_w}$$

5. Determine β from curve.
6. Calculate d_{max}:

$$d_{max} = \frac{(h_o)_{max}}{\beta}$$

Note: d_{max} depends on Q_o, Q_w, $(t_r)_o$, and $(t_r)_w$.

7. Calculate combinations of d, L_{eff} for d less than d_{max} that satisfy the gas capacity constraint. Use 100-μm droplet if no other information is available.

Field units

$$dL_{eff} = 420\left(\frac{TZQ_g}{P}\right)\left[\left(\frac{\rho_g}{\rho_1 - \rho_g}\right)\frac{C_D}{d_m}\right]^{1/2}$$

SI units

$$dL_{eff} = 34.5\left(\frac{TZQ_g}{P}\right)\left[\left(\frac{\rho_g}{\rho_1 - \rho_g}\right)\frac{C_D}{d_m}\right]^{1/2}$$

8. Calculate combinations of d, L_{eff} for d less than d_{max} that satisfy the oil and water retention time constraints.

Field units

$$d^2L_{eff} = 1.42[(t_r)_o Q_o + (t_r)_w Q_w]$$

SI units

$$d^2L_{eff} = 4.2 \times 10^4[(t_r)_o Q_o + (t_r)_w Q_w]$$

9. Estimate seam-to-seam length.

Field Units

$$L_{ss} = L_{eff} + \frac{d}{12} \quad \text{(gas capacity)}$$

$$L_{ss} = \frac{4}{3} L_{eff} \quad \text{(liquid capacity)}$$

SI units

$$L_{ss} = L_{eff} + \frac{d}{1000} \quad \text{(gas capacity)}$$

$$L_{ss} = \frac{4}{3} L_{eff} \quad \text{(liquid capacity)}$$

10. Select a reasonable diameter and length. Slenderness ratios ($12\, L_{ss}/d$) on the order of 3–5 are common.
11. When making a final selection, it is always more economical to select a standard vessel size. API sizes for small separators can be found in API Spec. 12J. In larger sizes in most locations, heads come in outside diameters, which are multiples of 6 in. (150 mm). The width of steel sheets for the shells is usually 10 ft (3000 mm), thus it's common practice to specify L_{ss} in multiples of five.

4.9.4 Horizontal Separators Sizing Other than Half-Full

For three-phase separators other than 50% full of liquid, equations can be derived similarly, using the actual oil and water areas. The equations are derived using the same principles as discussed in Chapter 3 and this chapter.

$$dL_{eff} = 420\left(\frac{1-\beta}{1-\alpha}\right)\left[\frac{TZQ_g}{P}\right]\left[\left(\frac{\rho_g}{\rho_l - \rho_g}\right)\frac{C_D}{d_m}\right]^{1/2} \tag{4.14a}$$

where $\dfrac{1-\beta}{1-\alpha}$ = design constant found from Figure 4.21.

SI units

$$dL_{eff} = 34.5\left(\frac{1-\beta}{1-\alpha}\right)\left[\frac{TZQ_g}{P}\right]\left[\left(\frac{\rho_g}{\rho_l - \rho_g}\right)\frac{C_D}{d_m}\right]^{1/2} \tag{4.14b}$$

where $\dfrac{1-\beta}{1-\alpha}$ = design constant found from Figure 4.21

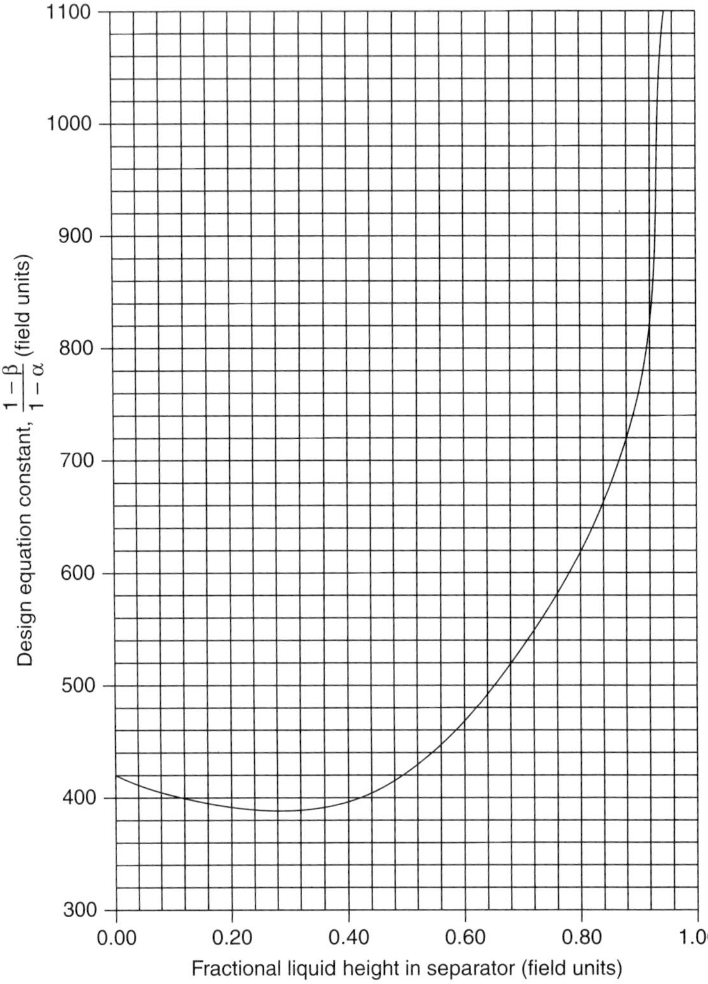

FIGURE 4.21. Gas capacity constraint design constant versus liquid height of a cylinder for a horizontal separator other than 50% full of liquid.

4.9.5 Gas Capacity Constraint (Figures 4.21 and 4.22)

Retention Time Constraint

Field units

$$d^2 L_{\mathrm{eff}} = \frac{(t_r)_o Q_o + (t_r)_w Q_w}{1.4\alpha} \qquad (4.15a)$$

where a = design constant found in Figure 4.22.

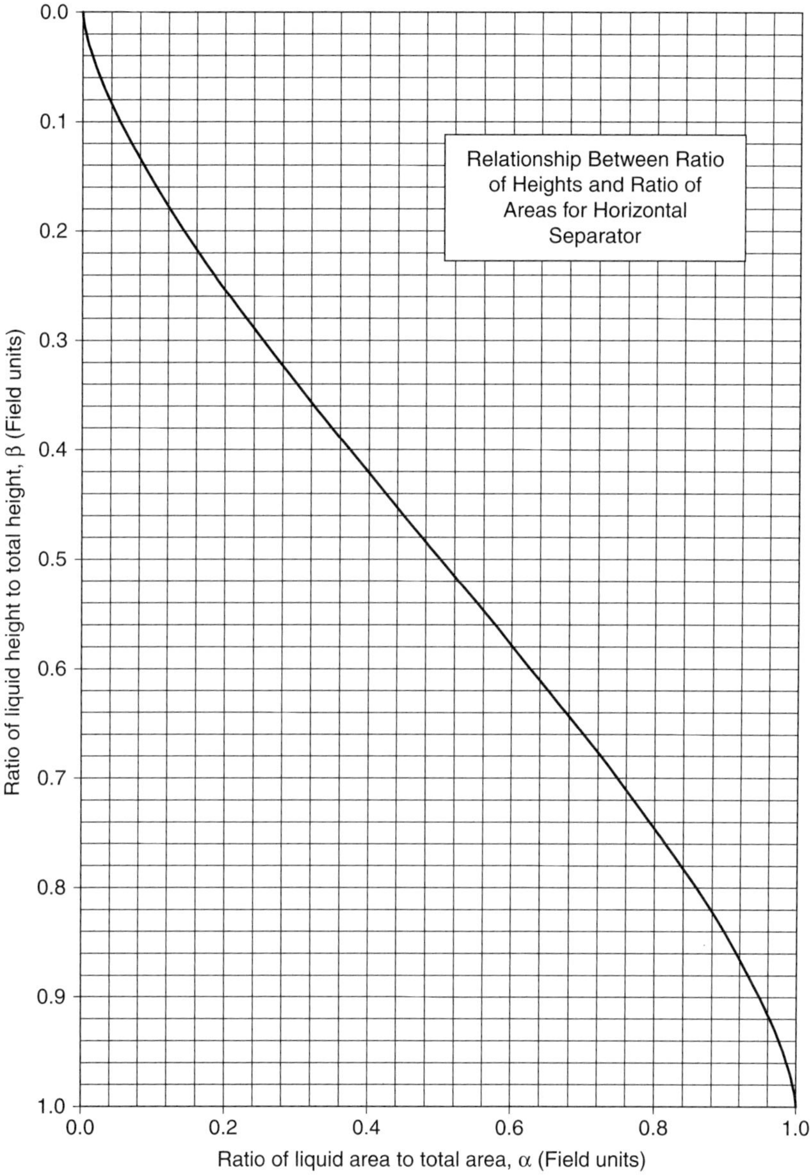

FIGURE 4.22. Retention time constraint design constant — ratio of areas (α) versus ratio of heights (β) for a horizontal separator other than 50% full of liquid.

SI units

$$d^2 L_{eff} = 21.000 \frac{(t_r)_o Q_o + (t_r)_w Q_w}{\alpha} \tag{4.15b}$$

where a = design constant found in Figure 4.22.

4.9.6 Settling Equation Constraint

From the maximum oil pad thickness, liquid flow rates, and retention times, a maximum vessel diameter may be calculated. The fractional cross-sectional area of the vessel required for water retention may be determined as follows:

$$\alpha_w = \frac{\alpha_1 Q_w (t_r)_w}{Q_o (t_r)_o + Q_w (t_r)_w} \tag{4.16}$$

where α_l = fractional area of liquids, α_w = fractional area of water.

The fractional height of the vessel required for the water can be determined by solving the following equation by trial and error:

$$\alpha_w = \frac{1}{80} \cos^{-1}[1 - 2\beta_w] - \left(\frac{1}{\pi}\right)[1 - 2\beta_w] \tag{4.17}$$

where β_w represents the fractional height of water.

A maximum vessel diameter may be determined from the fractional heights of the total liquids and water as follows:

$$d_{max} = ((h_o)_{max})/(\beta_1 - \beta_w) \tag{4.18}$$

where d_{max} is the maximum vessel internal diameter in inches (mm).

Any vessel diameter less than this maximum may be used to separate specified water droplet size in the specified oil retention time.

4.10 Vertical Separators' Sizing

As with vertical two-phase separators, a minimum diameter must be maintained to allow liquid droplets to separate from the vertically moving gas. The vessel must also have a large enough diameter to allow water droplets to settle in the upward-flowing oil phase and to allow oil droplets to rise in the downward-moving water phase. The liquid retention time requirement specifies a combination of diameter and liquid volume height. Any diameter greater than the minimum required for gas capacity and for liquid separation can be chosen.

4.10.1 Gas Capacity Constraint

By setting the gas velocity equal to the terminal settling velocity of a droplet, the following may be derived:

Field units

$$d^2 = 5040\left[\frac{TZQ_g}{P}\right]\left[\left(\frac{\rho_g}{\rho_1 - \rho_g}\right)\frac{C_D}{d_m}\right]^{1/2} \tag{4.19a}$$

SI units

$$d^2 = 34,500\left[\frac{TZQ_g}{P}\right]\left[\left(\frac{\rho_g}{\rho_1 - \rho_g}\right)\frac{C_D}{d_m}\right]^{1/2} \tag{4.19b}$$

For 100-μm droplet removal, Equations (5.19a) and (5.19b) are reduced to the following:

Field units

$$d^2 = 504\left[\frac{TZQ_g}{P}\right]\left[\left(\frac{\rho_g}{\rho_1 - \rho_g}\right)C_D\right]^{1/2} \tag{4.20a}$$

SI units

$$d^2 = 3450\left[\frac{TZQ_g}{P}\right]\left[\left(\frac{\rho_g}{\rho_1 - \rho_g}\right)C_D\right]^{1/2} \tag{4.20b}$$

4.10.2 Settling Water Droplets from Oil Phase

The requirement for settling water droplets from the oil requires that the following equation must be satisfied:

Field units

$$d^2 = 6,690\frac{Q_o\mu}{(\Delta SG)d_m^2} \tag{4.21a}$$

SI units

$$d^2 = 6.37 \times 10^8\frac{Q_o\mu}{(\Delta SG)d_m^2} \tag{4.21b}$$

For 500-μm droplets, Equations (4.21a) and (4.21b) become

Field Units

$$d^2 = 0.0267\left(\frac{Q_o\mu}{\Delta SG}\right) \tag{4.22a}$$

SI units

$$d^2 = 2550\left(\frac{Q_o\mu}{\Delta SG}\right) \tag{4.22b}$$

4.10.3 Settling Oil from Water Phase

The requirement for separating oil from water requires that the following equation must be satisfied:

Field units

$$d^2 = 6,690\left(\frac{Q_o\mu}{(\Delta SG)d_m^2}\right) \tag{4.21a}$$

SI units

$$d^2 = 6.37 \times 10^8\left(\frac{Q_o\mu}{(\Delta SG)d_m^2}\right) \tag{4.21b}$$

For 200-μm droplets, Equations (4.21a) and (4.21b) become

Field units

$$d^2 = 0.167\left(\frac{Q_o\mu}{(\Delta SG)}\right) \tag{4.23a}$$

SI units

$$d^2 = 1.59 \times 10^4\left(\frac{Q_o\mu}{(\Delta SG)}\right) \tag{4.23b}$$

4.10.4 Retention Time Constraint

Field units

$$h_o + h_w = \frac{[(t_r)_o Q_o + (t_r)_w Q_w]}{0.12d^2} \tag{4.24a}$$

SI units

$$h_o + h_w = \frac{[(t_r)_o Q_o + (t_r)_w Q_w]}{4.713 \times 10^{-8} d^2} \tag{4.24b}$$

where h_o = height of oil pad, in. (mm), h_w = height from water outlet to interface, in. (mm). (Note: this height must be adjusted for cone bottom vessels.)

4.10.5 Seam-to-Seam Length

As with horizontal three-phase separators, the specific design of the vessel internals will affect the seam-to-seam length. The seam-to-seam length (L_{ss}) of vertical vessels may be estimated based on the diameter and liquid height. As shown in Figure 4.23, allowance must be made for the gravity settling (gas separation) section, inlet diverter, mist extractor, and any space below the water outlet. For screening purposes, the larger L_{ss} values from Equations (4.25a and 4.25b) and (4.26a and 4.26b) should be used.

Field units

$$L_{ss} = \frac{h_o + h_w + 76}{12} \quad \text{(for diameters} \leq 36 \text{ in.).} \tag{4.25a}$$

$$L_{ss} = \frac{h_o + h_w + d + 40}{12} \quad \text{(for diameters} > 36 \text{ in.).} \tag{4.26a}$$

SI units

$$L_{ss} = \frac{h_o + h_w + 1930}{1000} \quad \text{(for diameters} \leq 914 \text{ mm),} \tag{4.25b}$$

$$L_{ss} = \frac{h_o + h_w + d + 1016}{1000} \quad \text{(for diameters} > 914 \text{ mm)} \tag{4.26b}$$

Where h_o = height of oil pad, in. (mm), h_w = height from water outlet to interface, in. (mm), d = vessel's internal diameter, in. (mm).

The larger of the L_{ss} values from Equations (4.25a) and (4.25b) as well as (4.26a) and (4.26b) should be used.

4.10.6 Slenderness Ratio

As with horizontal three-phase separators, the larger the slenderness ratio, the less expensive the vessel. In vertical separators whose sizing is liquid dominated, it is common to choose slenderness ratios no greater than 4 to keep the height of the liquid collection section to a reasonable level. Choices between 1.5 and 3 are common, although height restrictions may force the choice of a lower slenderness ratio.

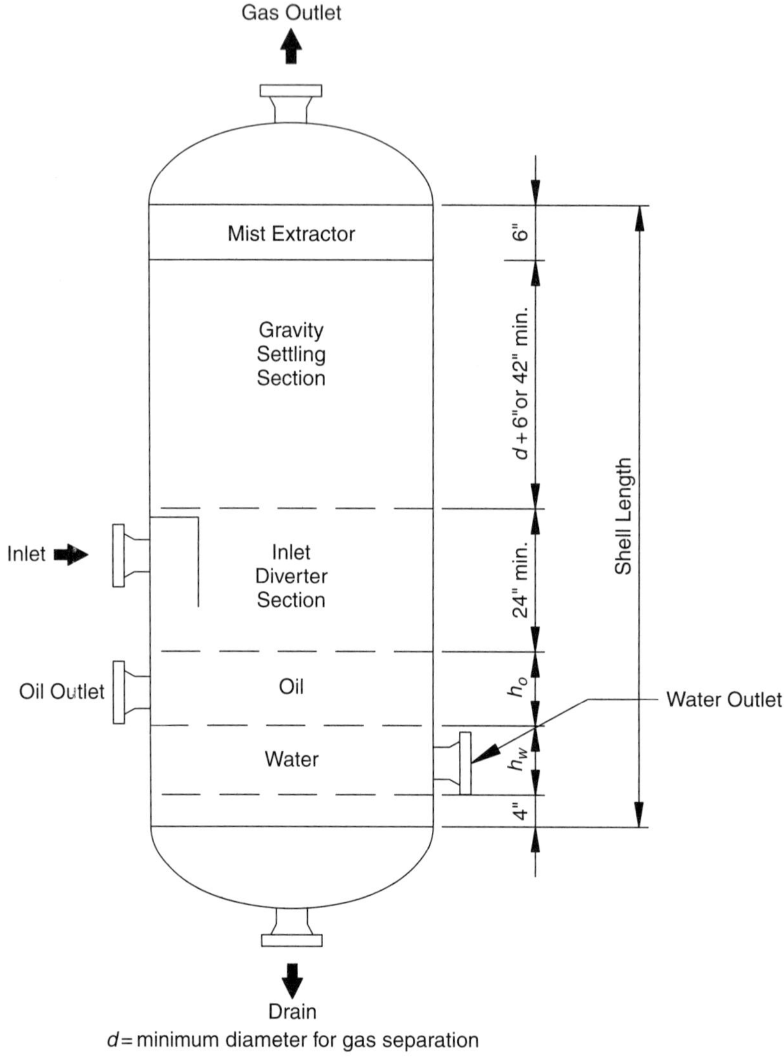

FIGURE 4.23. Approximate seam–seam shell length for a vertical three-phase separator.

4.10.7 Procedure for Sizing Three-Phase Vertical Separators

1. The first step in sizing a vertical separator is to establish the design basis. This includes specifying the maximum and minimum flow rates, operating pressure and temperature, droplet size to be removed, etc.
2. Equations (4.19a) and (4.19b) may be used to calculate the minimum diameter for a liquid droplet to fall through the gas phase.

Use Equations (4.20a) and (4.20b) for 100-μm droplets if no other information is available.

Field units

$$d^2 = 5040\left[\frac{TZQ_g}{P}\right]\left[\left(\frac{\rho_g}{\rho_l - \rho_g}\right)\frac{C_D}{d_m}\right]^{1/2} \qquad (4.19a)$$

SI units

$$d^2 = 34,500\left[\frac{TZQ_g}{P}\right]\left[\left(\frac{\rho_g}{\rho_l - \rho_g}\right)\frac{C_D}{d_m}\right]^{1/2} \qquad (4.19b)$$

For 100 μm:

Field units

$$d^2 = 504\left[\frac{TZQ_g}{P}\right]\left[\left(\frac{\rho_g}{\rho_l - \rho_g}\right)C_D\right]^{1/2} \qquad (4.20a)$$

SI units

$$d^2 = 3500\left[\frac{TZQ_g}{P}\right]\left[\left(\frac{\rho_g}{\rho_l - \rho_g}\right)C_D\right]^{1/2} \qquad (4.20b)$$

3. Equations (4.21a) and (4.21b) may be used to calculate the minimum diameter for water droplets to fall through the oil phase. Use Equations (4.22a) and (4.22b) for 500-μm droplets if no other information is available.

Field units

$$d^2 = 6690\frac{Q_o\mu}{(\Delta SG)d_m^2} \qquad (4.21a)$$

SI units

$$d^2 = 6.37 \times 10^8\frac{Q_o\mu}{(\Delta SG)d_m^2} \qquad (4.21b)$$

For 500-μm droplets:

Field units

$$d^2 = (0.0267)\left(\frac{Q_o\mu}{\Delta SG}\right) \qquad (4.22a)$$

SI units

$$d^2 = 2550\left(\frac{Q_o\mu}{\Delta SG}\right) \quad (4.22b)$$

4. Equations (4.21a) and (4.21b) may be used to calculate the minimum diameter for oil droplets to rise through the water phase. Use Equations (4.23a) and (4.23b) for 200-μm droplets if no other information is available.

For 200-μm droplets:

Field units

$$d^2 = 0.167\left(\frac{Q_o\mu}{(\Delta SG)}\right) \quad (4.23a)$$

SI units

$$d^2 = 1.59 \times 10^4\left(\frac{Q_o\mu}{(\Delta SG)}\right) \quad (4.23b)$$

5. Select the largest of the three diameters calculated in steps 2–4 as the minimum diameter. Any value larger than this minimum may be used for the vessel diameter.
6. For the selected diameter, and assumed values of $(t_r)_o$ and $(t_r)_w$, Equations (4.24a) and (4.24b) may be used to determine h_o+h_w

Field units

$$h_o + h_w = \frac{[(t_r)_o Q_o + (t_r)_w Q_w]}{0.12d^2} \quad (4.24a)$$

SI units

$$h_o + h_w = \frac{[(t_r)_o Q_o + (t_r)_w Q_w]}{4.713 \times 10^{-8} d^2} \quad (4.24b)$$

7. From d and h_o+h_w the seam-to-seam length may be estimated using Equations (4.25a and 4.25b) and (4.26a and 4.26b). The larger value of L_{ss} should be used.

Field units

$$L_{ss} = \frac{h_o + h_w + 76}{12} \quad \text{(for diameters} \leq 36\text{ in.)} \quad (4.25a)$$

SI units

$$L_{ss} = \frac{h_o + h_w + 1930}{1000} \quad \text{(for diameters} \leq 914\ \text{mm)} \tag{4.25b}$$

Field units

$$L_{ss} = \frac{h_o + h_w + d + 40}{12} \quad \text{(for diameters} > 36\ \text{in.)} \tag{4.26a}$$

SI units

$$L_{ss} = \frac{h_o + h_w + d + 1016}{1000} \quad \text{(for diameters} > 914\ \text{mm)} \tag{4.26b}$$

8. Check the slenderness ratios. Slenderness ratios between 1.5 and 3 are common. The following equations may be used:

Field units

$$\text{SR} = \frac{12L_{ss}}{d} \tag{4.27a}$$

SI units

$$\text{SR} = \left(\frac{L_{ss}}{(1000)d}\right) \tag{4.27b}$$

9. If possible, select a standard-size diameter and seam-to-seam length.

Examples

Example 4.1: sizing a vertical three-phase separator (field units)
Given

Q_o = 5000 BOPD,
Q_w = 3000 BWPD,
Q_g = 5 MMscfd,
P_o = 100 psia,
T_o = 90 °F,
Oil = 30 °API,
$(SG)_w$ = 1.07,
S_g = 0.6,
$(t_r)_o = (t_r)_w$ = 10 min,

$\mu_o = 10$ cp,
$\mu_w = 1$ cp,
$C_D = 2.01$

Droplet removal = 100 μm liquids, 500 μm water, 200 μm oil.

Solution

1. Calculate difference in specific gravities.

$$^\circ\text{API} = \frac{141.5}{(\text{SG})_o} - 131.5$$
$$= 0.876,$$
$$\Delta\text{SG} = 1.07 - 0.876 = 0.194$$

2. Calculate the minimum diameter required to settle a liquid droplet through the gas phase [Equation (4.19a)].

$$d^2 = 5040\left[\frac{(550)(0.99)(5)}{(100)}\right]\left[\left(\frac{0.3}{(54.7)-(0.3)}\right)\frac{2.01}{100}\right]^{1/2},$$
$$d = 34.9 \text{ in.}$$

3. Calculate the minimum diameter required for water droplets to settle through the oil phase [Equation (4.21a)].

$$d^2 = 6,690\left[\frac{Q_o\mu}{(\Delta\text{SG})d_m^2}\right]$$
$$= 6,690\left[\frac{(5,000)(10)}{(0.194)(500)^2}\right],$$
$$d = 83.0 \text{ in.}$$

4. Calculate the minimum diameter required for oil droplets to rise through the water phase [Equation (4.23a)].

$$d^2 = 6690\left[\frac{Q_o\mu}{(\Delta\text{SG})d_m^2}\right]$$
$$= 6690\left[\frac{(3000)(1)}{(0.194)(200)^2}\right],$$
$$d = 50.8 \text{ in.}$$

5. Select the largest diameter from steps 2–4 as the minimum inside diameter required.

$$d_{min} = 83.0 \text{ in.}$$

6. Calculate $h_o + h_w$.

$$h_o + h_w = \frac{(t_r)_o(Q_o) + (t_r)_w Q_w}{0.12d^2},$$

$$h_o + h_w = \frac{(10)(5000 + 3,000)}{0.12d^2}$$

$$= \frac{667,000}{d^2}$$

Refer to Table 4.2 for results.

TABLE 4.2
Vertical three-phase separator capacity diameter vs. length for retention time constraint $(t_r)_o = (t_r)_w = 10$ min

d_o (in.)	h_o+h_w (in.)	L_{ss} (ft)	$SR\left(\frac{12L_{ss}}{d_o}\right)$
84	94.5	18.2	2.6
90	82.3	17.7	2.4
96	72.3	17.4	2.2
102	64.1	17.2	2.0

7. Compute seam-to-seam length (L_{ss}). Select the larger value from Equation (4.25a) or (4.26a).

$$L_{ss} = \frac{h_o + h_w + 76}{12} \quad \text{(for diameters} \leq 36 \text{ in.)},$$

$$L_{ss} = \frac{h_o + h_w + d + 40}{12} \quad \text{(for diameters} > 36 \text{ in.)}$$

Refer to Table 4.2 for results.

8. Compute the slenderness ratio.

$$\text{Slenderness ratio} = \frac{12L_{ss}}{d}$$

Choices in the range of 1.5–3 are common. Refer to Table 4.2 for results.

9. Make final selection: compute combinations of d and h_o+h_w for diameters greater than the minimum diameter. See Table 4.2 for results. Select 90 in outside diameter (OD) × 20 ft seam-to-seam length (s/s).

Example 4.2: sizing a horizontal three-phase separator (field units)

Given

Q_o = 5000BOPD,
Q_w = 3000BWPD,
Q_g = 5MMscfd,
P = 100psia,
T = 90°F,
Oil = 30° API,
$(SG)_w$ = 1.07,
S_g = 0.6,
$(t_r)_o = (t_r)_w$ = 10 min,
μ_o = 10 cp,
μ_w = 1 cp,

Droplet removal = 100 μm liquid, 500 μm water, 200 μm oil. Vessel is half-full of liquids.

Solution

1. Calculate difference in specific gravities.

$$°\text{API} = \frac{141.5}{(\text{SG})_o} - 131.5$$

$$(\text{SG})_o = \frac{141.5}{30 + 131.5} = 0.876,$$

$$\Delta\text{SG} = 1.07 - 0.876 = 0.194$$

2. Calculate maximum oil pad thickness $(h_o)_{max}$. Use 500-micron droplet size if no other information is available.

$$(h_o)_{max} = (1.28 \times 10^{-3}) \frac{(t_r)_o(\Delta SG)d_m^2}{\mu}$$

$$= 0.00128 \frac{(10)(0.194)(500)^2}{10}$$

$$= 62.1$$

3. Calculate $\frac{A_w}{A}$:

$$\frac{A_w}{A} = 0.5\frac{Q_w(t_r)_w}{(t_r)Q_o + (t_r)_w Q_w}$$

$$= 0.5\frac{(19.8)(10)}{(33)(10) + (19.8)(10)}$$

$$= 0.1875$$

4. Determine β from Figure 4.20. with $A_w/A = 0.1875$, read $\beta = 0.257$.

5. Calculate d_{max}.

$$d_{max} = \frac{(h_o)_{max}}{\beta}$$

$$= \frac{62.1}{0.257},$$

$$d_{max} = 241.6 \text{ in.}$$

6. Calculate combinations of d, L_{eff} for d less than d_{max} that satisfy the gas capacity constraint. Use 100-μm droplet size if no other information is available.

$$dL_{eff} = 420\left(\frac{TZQ_g}{P}\right)\left[\left(\frac{\rho_g}{\rho_l - \rho_g}\right)\frac{C_D}{d_m}\right]^{1/2}$$

$$= 420\left(\left(\frac{(550)(0.99)(5)}{(100)}\right)\left(\frac{0.3}{(54.7) - (0.3)}\right)\frac{2.01}{100}\right)^{1/2}$$

$$= 120$$

Refer to Table 4.3 for results.

TABLE 4.3
Horizontal three-phase separator diameter vs. length for gas capacity constraint

d (in.)	L_{eff} (ft)
60	1.7
72	1.4
84	1.2
96	1.1

Since the values of L_{eff} are low, the gas capacity does not govern.

7. Calculate combinations of d, L_{eff} for d less than $d_{\max}$ that satisfy the oil and water retention time constraints.

$$
\begin{aligned}
d^2 L_{\mathrm{eff}} &= 1.42[Q_{\mathrm{w}}(t_{\mathrm{r}})_{\mathrm{w}} + Q_{\mathrm{o}}(t_{\mathrm{r}})_{\mathrm{o}}] \\
&= (1.42)(10)(8,000) \\
&= 113,600
\end{aligned}
$$

Refer to Table 4.4 for results.

TABLE 4.4
Horizontal three-phase separator capacity diameter vs. length for liquid retention time constraint $(t_r)_o$=$(t_r)_w$ 10 mm

d (In.)	L_{eff} (ft)	L_{ss} (ft)	$\mathrm{SR}\left(\frac{12L_{\mathrm{ss}}}{d}\right)$
60	31.6	42.1	8.4
72	21.9	29.2	4.9
84	16.1	21.5	3.1
96	12.3	16.4	2.1
108	9.7	13.0	1.4

8. Estimate seam-to-seam length.

$$L_{\mathrm{ss}} = L_{\mathrm{eff}} + \frac{d}{12} \quad \text{(for gas capacity)},$$

$$L_{\mathrm{ss}} = \frac{4}{3} L_{\mathrm{eff}} \quad \text{(for liquid capacity)}.$$

9. Select slenderness ratio (12 L_{ss}/d). Choices in the range of 3–5 are common.
10. Choose a reasonable size that does not violate gas capacity restraint or oil pad thickness restraint. Possible choices are 72 in. diameter by 30 ft seam-by-seam and 84 in. diameter by 25 ft seam-by-seam.

Nomenclature

A_{l} cross-sectional area of vessel available for liquid retention, ft^2 (m^2)

A_{T} total cross-sectional area of vessel, ft^2 (m^2)

A_w	cross-sectional area of vessel available for water retention, ft^2 (m^2)
API	API gravity of oil, °API
C_D	drag coefficient, dimensionless
D_m	drop diameter, ft(m)
D	vessel internal diameter, ft (m)
d	vessel internal diameter, in. (mm)
d_l	water leg standpipe internal diameter, in. (mm)
d_m	drop diameter, μm
d_{max}	maximum vessel internal diameter, in. (mm)
d_{min}	minimum allowable vessel internal diameter, in. (mm)
d_o	vessel external diameter, in. (mm)
F_B	buoyant force, lb (N)
F_D	drag force, lb (N)
g	gravitational constant, $32.2 lb_m ft/lb_f s^2$ ($9.81\ m/s^2$)
H	height of liquid volume, ft (m)
h	height of liquid volume, in. (mm)
H_l	height of liquid in horizontal vessel, ft (m)
h_l	height of liquid in horizontal vessel, in. (mm)
H_o	height of oil pad, ft (m)
h_o	height of oil pad, in (mm)
$(h_o)_{max}$	maximum oil pad thickness, in. (mm)
H_w	height from water outlet to interface, ft (m)
h_w	height from water outlet to interface, in. (mm)
$(h_w)_{max}$	maximum water height, in. (mm)
h'_w	height of water weir, in. (mm)
L_{eff}	effective length of the vessel, ft (m)
L_{ss}	vessel length seam-to-seam, ft (m)
P	operating pressure, psia (kPa)
Q	flow rate, ft^3/s (m^3/s)
Q_g	gas flow rate, MMscfd (std m^3/h)
Q_l	liquid flow rate, BPD (m^3/h)
Q_o	oil flow rate, BPD (m^3/h)
Q_w	water flow rate, BPD (m^3/h)
SR	Slenderness ratio, dimensionless
SG	oil specific gravity
T	operating temperature, °R (K)
T	temperature, °F (°C)
t_d	droplet settling time, s
t_g	gas retention time, s
t_o	oil retention time or settling time, s
t_r	liquid retention time, min
$(t_r)_o$	oil retention time, min
$(t_r)_w$	water retention time, min
t_w	water retention time or settling time, s

V	volume, ft^3 (m^3)
V_g	gas velocity, ft/s (m/s)
$(V_g)_{max}$	maximum gas velocity, no re-entrainment, ft/s (m/s)
V_l	average liquid velocity, ft/s (m/s)
V_o	oil volume, ft^3 (m^3)
V_t	terminal settling velocity of the droplet, ft/s (m/s)
V_w	water volume, ft^3 (m^3)
Z	gas compressibility factor, dimensionless
α	fractional cross-sectional area of liquid
α_l	fractional area of liquids
α_o	fractional area of oil
α_w	fractional area of water
β	fractional height of liquid within the vessel $= h_l/d_l$
β_l	fractional height of liquid
β_w	fractional height of water
Δ_h	height difference between oil weir and water weir, in. (mm)
ΔSG	difference in specific gravity relative to water of the drop and the gas
θ	angle used in determining α, radians or degrees
μ	viscosity of continuous phase, cp (Pa s)
μ_l	dynamic viscosity of the liquid, lb_m/ft-s (kg/m-s)
μ_o	viscosity of oil phase, cp (Pa s)
μ_w	viscosity of water phase, cp (Pa s)
ρ	density of the continuous phase, lb/ft^3 (kg/m^3)
ρ_g	density of the gas at the temperature and pressure in the separator, lb/ft^3 (kg/m^3)
ρ_l	density of liquid, lb/ft^3 (kg/m^3)
ρ_o	oil density, lb/ft^3 (kg/m^3)
ρ_w	water density, lb/ft^3 (kg/m^3)

CHAPTER 5

Mechanical Design of Pressure Vessels

5.1 Introduction

Chapters 3 and 4 discuss the concepts for determining the diameter and length of two-phase and three-phase vertical and horizontal separators. This chapter addresses the selection of design pressure rating and wall thickness of pressure vessels. It also presents a procedure for estimating vessel weight and includes some examples of design details.

The purpose of this chapter is to present an overview of simple concepts of mechanical design of pressure vessels that must be understood by a project engineer specifying and purchasing this equipment. Most pressure vessels used in the oil and gas industry are designed and inspected according to the American Society of Mechanical Engineers' Boiler and Pressure Vessel Code (ASME code). Because the ASME code contains much more detail than can be covered in a single chapter of a general textbook such as this one, the project engineer should have access to a copy of the ASME code and should become familiar with its general contents. In particular, Section VIII of the code, "Pressure Vessels," is particularly important. Countries that do not use the ASME code have similar documents and requirements. The procedures used in this chapter that refer specifically to the ASME code are generally applicable in other countries but should be checked against the applicable code.

In U.S. federal waters and the majority of countries with oil and gas operations, all pressure vessels must be designed and inspected in accordance with the ASME code. In some countries, however, there is no such requirement. It is possible to purchase "noncode" vessels in these countries at a small savings in cost. Non-code vessels are normally designed to code requirements (although there is no certainty that this is true), but they are not inspected by a qualified code inspector nor are they necessarily inspected to the quality standards dictated by the code. For this reason, the use of noncode vessels should be discouraged to assure vessel mechanical integrity.

5.2 Design Considerations

5.2.1 Design Temperature

The maximum and minimum design temperatures for a vessel will determine the maximum allowable stress value permitted for the material to be used in the fabrication of the vessel. The maximum temperature used in the design should not be less than the mean metal temperature expected under the design operating conditions. The minimum temperature used in the design should be the lowest expected in service except when lower temperatures are permitted by the rules of the ASME code.

In determining the minimum temperature, such factors as the lowest operating temperature, operational upset, auto-refrigeration, ambient temperature, and any other source of cooling should all be considered. If necessary, the metal temperature should be determined by computation using accepted heat transfer procedures or by measurement from equipment in service under equivalent operating conditions.

5.2.2 Design Pressure

The design pressure for a vessel is called its maximum allowable working pressure (MAWP). In conversation this is sometimes referred to simply as the vessel's working pressure. The MAWP determines the setting of the relief valve and must be higher than the normal pressure of the process contained in the vessel, which is called the vessel's operating pressure. The operating pressure is fixed by process conditions. Table 5.1 recommends a minimum differential between operating pressure and MAWP so that the difference between the operating pressure and the relief valve set pressure provides a sufficient cushion. If the operating pressure is too close to the relief valve setting, small surges in operating pressure could cause the relief valve to activate prematurely.

Some vessels have pressure safety high sensors (PSHs) that shut in the inflow if a higher-than-normal pressure is detected. The use of PSHs is discussed in more detail in the Instrumentation, Process Control and Safety Systems volume of this series. The differential between the maximum operating pressure and the PSH sensor set pressure should be as indicated in Table 5.1, and the relief valve

TABLE 5.1
Setting maximum allowable working pressures minimum differential between operating pressure

Operating Pressure and MAWP	
Less than 50 psig	10 psi to 51–250 psig
25 psi to 251–500 psig	10% of maximum operating pressure 501–1000 psig
50 psi to 1001 psig and higher	5% of maximum operating pressure
Vessels with high-pressure safety sensors have an additional	5% or 5 psi, whichever is greater to the minimum differential

should be set at least 5% or 5 psi, whichever is greater, higher than the PSH sensor set pressure. Thus, the minimum recommended MAWP for a vessel operating at 75 psig with a PSH sensor would be 105 psig (75 + 25 + 5); the PSH sensor is set at 100 psig and the relief valve is set at 105 psig.

Often, especially for small vessels, it is advantageous to use a higher MAWP than is recommended in Table 5.1. It may be possible to increase the MAWP at little or no cost and thus have greater future flexibility if process changes (e.g., greater throughput) require an increase in operating pressure.

The MAWP of the vessel cannot exceed the MAWP of the nozzles, valves, and pipe connected to the vessel. As discussed in the Plant Piping and Pipeline volume of this series, pipe flanges, fittings, and valves are manufactured in accordance with industry standard pressure rating classes. Table 5.2 is a summary of Material Group 1.1 carbon steel fittings manufactured in accordance with American National Standards Institute (ANSI) specification B16.5.

TABLE 5.2
Summary ANSI pressure ratings material Group 1.1

	MAWP (psig)	
Class	*–20 to 100 °F*	*F100 °F to 200 °F*
150	285	250
300	740	675
400	990	900
600	1480	1350
900	2220	2025
1500	3705	3375
2500	6170	5625

If the minimum MAWP calculated from Table 5.1 is close to one of the ANSI MAWP listed in Table 5.2, it is common to design the pressure vessel to the same MAWP as the ANSI class. For example, the 105-psig pressure vessel previously discussed will have nozzles, valves, and fittings attached to it that are rated for 285 psig (ANSI Class 150). The increase in cost of additional vessel wall thickness to meet a MAWP of 285 psig may be small.

Often, a slightly higher MAWP than that calculated from Table 5.1 is possible at almost no additional cost. Once a preliminary MAWP is selected from Table 5.1, it is necessary to calculate a wall thickness for the shell and heads of the pressure vessel. The procedure for doing this is described in the following section. The actual wall thickness chosen for the shell and heads will be somewhat higher than that calculated, as the shells and heads will be formed from readily available plates. Thus, once the actual wall thickness is determined, a new MAWP can be specified for essentially no additional cost. (There will be a marginal increase in cost to test the vessel to the slightly higher pressure.)

This concept can be especially significant for a low-pressure vessel where a minimum wall thickness is desired. For example, assume the calculations for a 50-psig MAWP vessel indicate a wall thickness of 0.20 in., and it is decided to use 0.25-in. plate. This same plate might be used if a MAWP of 83.3 psig were specified. Thus, by specifying the higher MAWP (83.3 psig), additional operating flexibility is available at essentially no increase in cost. Many operators specify the MAWP based on process conditions in their bids and ask the vessel manufacturers to state the maximum MAWP for which the vessel could be tested and approved.

5.2.3 Maximum Allowable Stress Values

The maximum allowable stress values to be used in the calculation of a vessel's wall thickness are given in the ASME code for many different materials. These stress values are a function of temperature. Section VIII of the ASME code, which governs the design and construction of all pressure vessels with operating pressures greater than 15 psig, is published in two divisions. Each sets its own maximum allowable stress values. Division 1, governing the *design by rules,* is less stringent from the standpoint of certain design details and inspection procedures, and thus incorporates a higher safety factor. The 1998 edition incorporates a safety factor of 4 while the 2001 and later editions incorporate a safety factor of 3.5.

The 2001 edition of the code yields higher allowable stresses and thus smaller wall thicknesses. For example, using a material with a 60,000-psi tensile strength, a vessel built under the 1998 edition (safety factor = 4) yields a maximum allowable stress value of 15,000 psi, while a vessel built under the 2001 edition (safety factor = 3.5) yields a maximum allowable stress value of 17,142 psi. On the other hand, Division 2 governs the *design by analysis* and incorporates a lower safety factor of 3. Thus, the maximum allowable stress value for a 60,000-psi tensile strength material will become 20,000 psi.

Many companies require that all their pressure vessels be constructed in accordance with Division 2 because of the more exacting standards. Others find that they can purchase less expensive vessels by allowing manufacturers the choice of either Division 1 or Division 2. Normally, manufacturers will choose Division 1 for low-pressure vessels and Division 2 for high-pressure vessels.

The maximum allowable stress values at normal temperature range for the steel plates most commonly used in the fabrication of pressure vessels are given in Table 5.3. For stress values at higher temperatures and for other materials, the latest edition of the ASME code should be referenced.

5.2.4 Determining Wall Thickness

The following formulas are used in the ASME code Section VIII, Division 1 for determining wall thickness:

Wall Thickness—Cylindrical Shells

$$t = \frac{Pr}{SE - 0.6P}, \tag{5.1}$$

Wall Thickness—2:1 Ellipsoidal Heads

$$t = \frac{Pd}{2SE - 0.2P}, \tag{5.2}$$

Wall thickness—Hemispherical Heads

$$t = \frac{Pr}{2SE - 0.2P}, \tag{5.3}$$

Wall Thickness—Cones

$$t = \frac{Pd}{2\cos\alpha\,(SE - 0.6P)}. \tag{5.4}$$

where S = maximum allowable stress value, psi (kPa), t = thickness, excluding corrosion allowance, in. (mm), P = maximum allowable working pressure, psig (kPa), r = inside radius before corrosion allowance is added, in. (mm), d = inside diameter before corrosion allowance is added, in. (mm), E = joint efficiency, see Table 5.4 (most vessels are fabricated in accordance with type of joint no. 1), a = half the angle of the apex of the cone.

TABLE 5.3
Maximum allowable stress value for common steels (2007 Edition)

			ASME Section VIII 2007 Edition	
			Div. 1	*Div. 2*
Metal		*Not Lower Than*	*−20 °F*	*−20 °F*
Temperature		*Not Exceeding*	*650 °F*	*100 °F*
Carbon steel plates and sheets	SA-516	Grade 55	15,700	18,300
		Grade 60	17,100	20,000
		Grade 65	18,600	21,700
		Grade 70	20,000	23,300
	SA-285	Grade A	12,900	15,000
		Grade B	14,300	16,700
		Grade C	15,700	18,300
	SA-36		16,600	16,900
Low-alloy steel plates	SA-387	Grade 2, cl.1	15,700	18,300
		Grade 12, cl.1	15,700	18,300
		Grade 11, cl.1	17,100	20,000
		Grade 22, cl.1	17,100	20,000
		Grade 21, cl.1	17,100	20,000
		Grade 5, cl.1	17,100	20,000
		Grade 2, cl.2	20,000	23,300
		Grade 12, cl.2	18,600	21,700
		Grade 11, cl.2	21,400	25,000
		Grade 22, cl.2	21,400	25,000
		Grade 21, cl.2	21,400	25,000
		Grade 5, cl.2	21,400	25,000
	SA-203	Grade A	18,600	21,700
		Grade B	20,000	23,300
		Grade D	18,600	21,700
		Grade E	20,000	23,300
High-alloy steel plates	SA-240	Grade 304	20,000	20,000**
		Grade 304L	16,700	16,700
		Grade 316	20,000	20,000
		Grade 316L	16,700	16,700

Austenitic stainless set at 2/3 yield/allowable stress, *not* 3.0 or 3.5 S.F due to low yield strength values relative to ultimate tensile strength, 304 UTS 75,000 Yield 30,000.
Example: Hydrostatic testing 1.3 × 20,000 = 26,000 (Yield is 30,000) for 304.

TABLE 5.4
Maximum allowable joint efficiencies for arc and gas welded joints

No.	Type of Joint Description	Limitation	Fully Radiographed[a]	Spot Examined[b]	Not Spot Examined[c]
1.	Butt joints as attained by double welding or by other means that will obtain the same quality of deposited weld metal on the inside and outside weld surfaces of UW-35. Welds using metal backing strips that remain in the place are excluded	None	1.00	0.85	0.70
2	Singled-welded butt joint with backing strip other than those include under (1)	(a) None except as in (b) below (b) Butt weld with one plate offset for circumferential joints only, see UW-13(c) and Figure UW-13.1(k)	0.90	0.80	0.65
3	Single-welded butt joint without use backing strip	Circumferential joints only, not over 5/8-in. thick and not over 24-in. outside diameter.	—	—	0.60
4	Double full filet lap joint	Longitudinal joints only, not over 3/8-in. thick.	—	—	0.55

(Continued)

TABLE 5.4 (*Continued*)

No.	*Type of Joint Description*	*Limitation*	*Fully Radiographed*[a]	*Spot Examined*[b]	*Not Spot Examined*[c]
5	Single full fillet lap joints with plug welds conforming to UW-17	(a) Circumferential joints[d] for attachment of heads not over 24-in. outside diameter to shells not over 0.5 in. thick. (b) Circumferential joints for the attachment to shells of jackets not over 5/8 in. in nominal thickness where the distance from the center of the plug weld to the edge of the plate is not less than 1.5 times the diameter of the hole for the plug.	—	—	0.50
6	Single full fillet lap joints with out plug welds	(a) For the attachment of heads convex to pressure to shells not over 5/8-in. required thickness, only with use of fillet weld on inside of shell; or (b) For attachment of heads having pressure on either side to shells not over 0.25-in. required thickness with fillet weld on out side of head flange only.	—	—	0.45

[a]*See UW-12(a) and UW-51.*
[b]*See UW-12(b) and UW-52.*
[c]*The maximum allowable joint efficiencies shown in this column are the weld joint efficiencies multiplied by 0.80 (and rounded off to the nearest 0.05) to effect the basic reduction in allowable stress required by the division for welded vessels that are sot spot examined. See (UW-12(c)).*
[d]*Joints attaching hemispherical heads to shells are executed.*

Figure 5.1 summarizes the formulas for pressure vessels under internal pressure (ASME Section VIII, Division 1). Figure 5.2 defines the various types of heads. Most production facility vessels use 2:1 ellipsoidal heads because they are readily available, are normally less expensive, and take up less room than hemispherical heads.

Cone-bottom vertical vessels are sometimes used where solids are anticipated to be a problem. Most cones have either a 90° apex

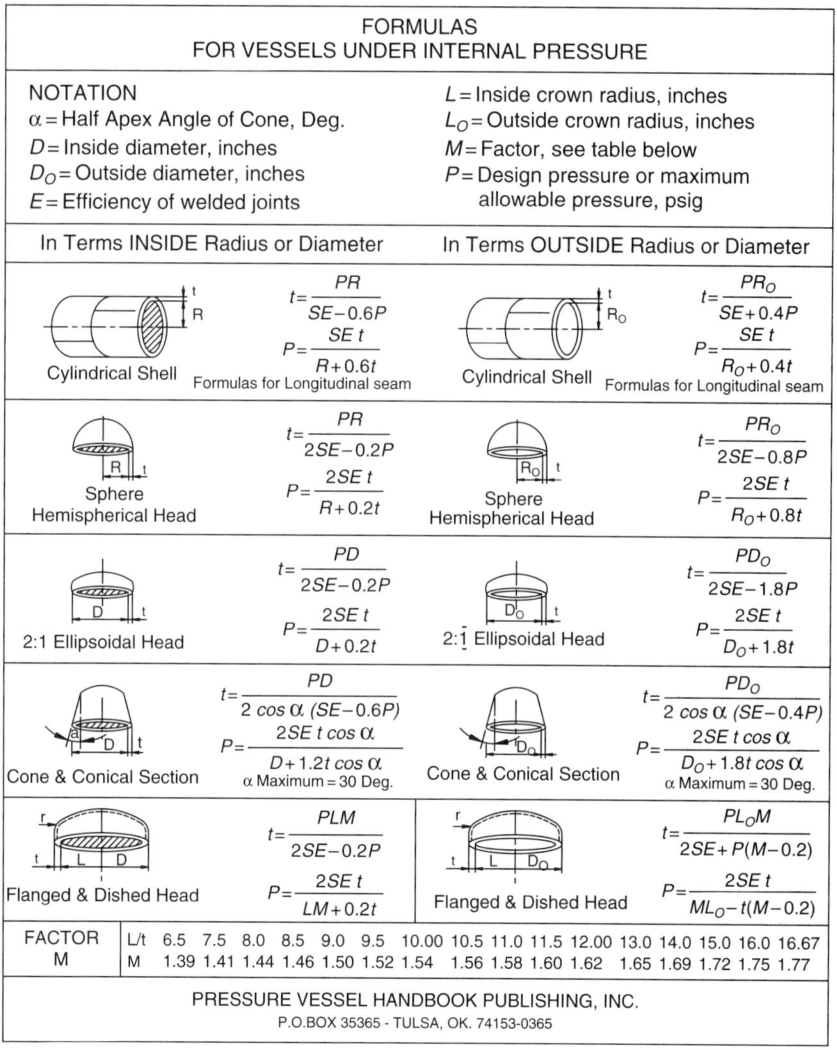

FIGURE 5.1. Formulas for vessels under internal pressure (ASME Section VIII, Division 1). (Reprinted with permission from Pressure Vessel Handbook, Publishing, Inc., Tulsa)

FORMULAS
FOR VESSELS UNDER INTERNAL PRESSURE

1. THE FORMULAS CONFORM TO THE ASME CODE FOR PRESSURE VESSELS. SECTION VIII, DIVISION L
2. CORROSION ALLOWANCE. If the vessel is subject to thinning by corrosion allowance shall be added to the thickness calculated by the formulas. Code UG-25.
   When allowance for corrosion is provided computing the wall thickness or allowable pressure in terms of the inside dimensions, the allowance shall be added lo the length of inside radius or diameter.
3. MAXIMUM ALLOWABLE WORKING PRESSURE for a vessel is the pressure allowable for the weakest element of the vessel. It should be limited not by minor parts but by the shell, heads of flanges.
4. HYDROSTATIC TEST PRESSURE o one and one-half times the maximum allowable working pressure or the design pressure when calculations are not made to determine the maximum allowable working pressure. Code UG-99.
5. DEGREE OF RADIOGRAPHIC EXAMINATION. Condition when full radiography is mandatory are described in the Code UW-11.
6. STRESS VALUES OF MATERIALS, S 1000 psi. Excerpt from Code, table UCS-73

Specification		For Metal Temperature Not Exceeding Deg. F.							
Number	Grade	-20 to 650	700	750	800	850	900	950	1010
SA-283	C	12.6	-	-	-	-		-	-
SA-285	C	13.7	13.2	12.0	10.2	8.3	6.5	-	-
SA-515	55	13.7	13.2	12.0	10.2	8.3	6.5	4.5	2.5
SA-515	60	15.0	14.3	12.9	10.8	8.6	6.5	4.5	2.5
SA-5I5	65	16.2	15.5	13.8	11.4	8.9	6 5	4 5	2.5
5A-515	70	17.5	16.6	14.7	12.0	9.2	6.5	4.2	2.5
SA-516	55	13.7	13.2	12.0	10.2	8.3	6.5	4.5	2.5
SA-516	60	15.0	14.3	12.9	10.8	8.6	6.5	4.5	2.5
SA-516	65	16.2	15.5	13.8	11.4	8.9	6 5	4.5	2.5
SA-516	70	17.5	16.6	14.7	12.0	9.2	6.5	4.5	2.5

7. CYLINDRICAL SHELL- In the formulas of the form the stress in the longitudinal joint are considered, since usually this governs.

   Stress in the girth seam will govern only when the circumferential joint efficiency is less than half the longitudinal joint efficiency, or when besides the internal pressure additional loadings (wind load, reaction of saddles etc.) arc outing longitudinal bending or tension.

   The formula considering the stress in the girth seam

$$t = \frac{PR}{2SE + 0.4P} \qquad P = \frac{2SE\,t}{R - 04t}$$

   When the thickness of the shell exceed one half of the inside radius, or P exceeds 0.385 SE. The formulas given in the Code UA-2(c) shall be applied.

8. SPHERE AND HEMISPHERICAL HEAD. When the wall thickness exceeds 0.356 R, or P exceeds 0.665 SE, the formula given in the Code UA-3 shall be applied.
9. ELLIPSOIDAL HEAD. When the ratio of the major and minor axis is other than 2:1. Formulas and factors given in the Code UA-4(c) shall be applied.
10. CONE AND CONICAL SECTION. The formulas given in the form are applicable only when the half apex angle of cone does not exceed 30 degree.
11. ASME FLANGED AND DISHED HEAD. The formulas are applicable when the inside crown radius is not greater than the outside diameter of the head and the inside knuckle radius is not less than 6% of the outside diameter or 3 times the thickness of the head.
12. EFFICIENCY OF WELDED JOINTS (E)
    Excerpt from ASME Code Table UW-12.

No.	TYPE OF JOINT	DEGREE OF EXAMINATION		
		FULL	SPOT	NO
1	Double-welded bolt joint or single welded bolt joint with backing strip which does not remain in place.	1.00	0.85	0.70
2	Single-welded bolt joint with backing strip which remains in place.	0.90	0.80	0.65
3	Single-welded bolt joint without use of backing strip.	---	---	0.60

13. EFFICIENCY (E) TO BE USED IN CALCULATIONS OF SEAMLESS HEADS ASME Code UW-12 (b)

TYPE OF HEAD	TYPE OF JOINT	DEGREE OF EXAMINATION OF HEAD TO SHELL JOINT		
		FULL	SPOT	NO
Hemi Spherical	No 1	1.00	0.85	0.80
	No 2	0.90	0.80	
Others	ANY	1.00	0.85	

14. NOTATION

a	=	Half-apex angle of cone, deg.
D	=	Inside diameter of cone at the large end, inches Inside diameter of ellipsoidal head, inches
D_0	=	Outside diameter of cone at the large end, inches Outside diameter of ellipsoidal head, inches
E	=	Lowest efficiency of any joint in the shell or head; for hemispherical head this includes head to shell joint
L	=	Inside crown radius of flanged and dished head, inches
L_0	=	Outside crown radius of flanged and dished head, inches
P	=	Internal design pressure or maximum allowable working pressure pounds per square inches
r	=	Knuckle radius
R	=	Inside radius of shell or hemispherical head, inches
R_0	=	Outside radius of shell or hemispherical head, inches
S	=	Allowable stress value of material pounds per square, inches
t	=	Thickness of shell or head, inches

FIGURE 5.1.—cont'd.

($a = 45°$) or a 60° apex ($a = 30°$). These are referred to respectively as a 45° or 60° cone because of the angle each makes with the horizontal. Equation (5.4) is for the thickness of a conical head that contains pressure.

Some operators use internal cones within vertical vessels with standard ellipsoidal heads as shown in Figure 5.3. The ellipsoidal heads contain the pressure, and thus the internal cone can be made of very thin steel.

Table 5.4 lists joint efficiencies that should be used in Equations (5.1)–(5.4). This is Table UW-12 in the ASME code. Table 5.5 lists some of the common material types used to construct pressure vessels. Individual operating companies have their own standards, which differ from those listed in this table.

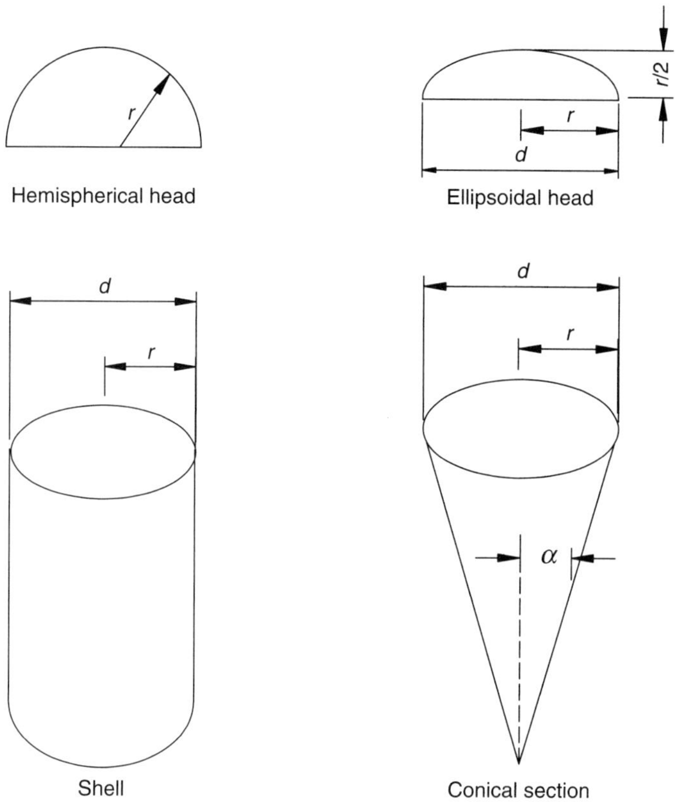

FIGURE 5.2. Pressure vessel shapes.

5.2.5 Corrosion Allowance

Typically, a corrosion allowance of 0.125 in. for non-corrosive service and 0.250 in. for corrosive service is added to the wall thickness calculated in Equations (5.1)–(5.4).

5.3 Inspection Procedures

All ASME code vessels are inspected by an approved code inspector. The manufacturer will supply code papers signed by the inspector. The name-plate on the vessel will be stamped to signify it has met the requirements of the code. One of these requirements is that the vessel was pressure tested (1998 edition, 1.5 times the MAWP; 2001 and later editions, 1.3 times the MAWP). However, this is only one of the requirements. The mere fact that a vessel is pressure tested 1.3 or

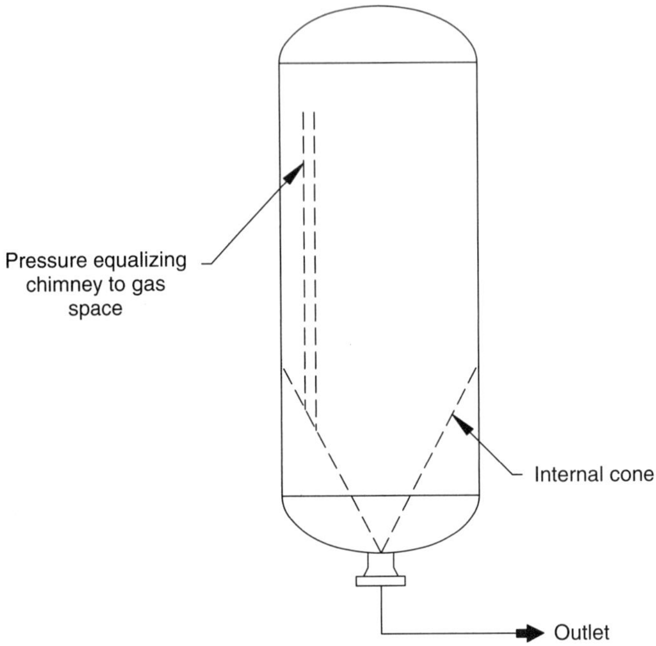

FIGURE 5.3. Internal cone vessel.

1.5 times the MAWP does not signify that it has met all the design and quality assurance safety aspects of the code.

It must be pointed out that a code stamp does not necessarily mean that the vessel is fabricated in accordance with critical nozzle dimensions or internal devices as required by the process. The code inspector is only interested in those aspects that relate to the pressure handling integrity of the vessel. The owner must do his own inspection to assure that nozzle locations are within tolerance, vessel internals are installed as designed, coatings are applied properly, and so on.

5.4 Estimating Vessel Weights

It is important to be able to estimate vessel weights, since most cost estimating procedures start with the weight of the vessel. The vessel weight, both empty and full with water, may be necessary to adequately design a foundation or to assure that the vessel can be lifted or erected once it gets to the construction site.

The weight of a vessel is made up of the weight of the shell, the weight of the heads, and the weight of internals, nozzles, pedestals, and skirts. The last two terms are defined in Figure 5.4.

TABLE 5.5
Materials typically specified

	Low Pressure	*Common Steel* T > −20 °F	*NACE MR-01-75*	*Low Temp −50 °F* < T < −0 °F	*Low Temp FT< −50 °F*	*High CO_2 Service*
Plate	SA-36, SA-285-C	SA-516-70	SA-516-70	SA-516-70	SA-240-304	SA-240-16L
Pipe	SA-53-B	SA-106-B	SA-106B	SA-106-B	SA-333-6, TP-304	SA-312, TP-316L
Flanges and Fittings	SA-105	SA-105, SA-181-1	SA-105, SA-181-1	SA-350-LF1	SA-182, F-304	SA-182, F-316L
Stud Bolts	SA-193-B7	SA-193-B7	SA-193-B7M	SA-320-L7	SA-193-B-8	SA-193-8M
Nuts	SA-192-2H	SA-194-2H	SA-194-2M	SA-194-4	SA-194-8A	SA-194-MA

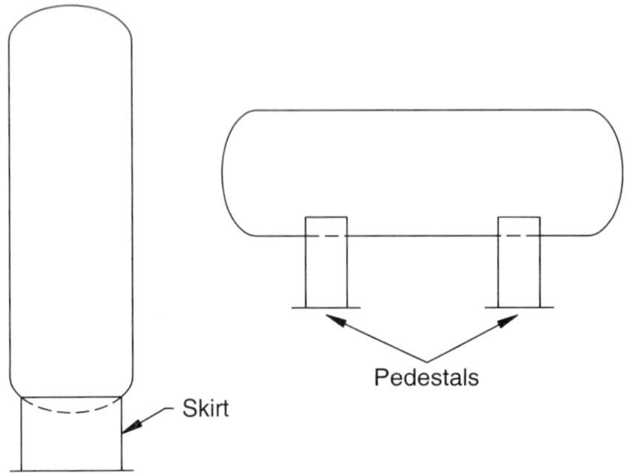

FIGURE 5.4. Vessel support devices.

The shell weight can be estimated from

Field units

$$W = 11\,dt\,L \tag{5.5a}$$

SI units

$$W = 0.0254\,dt\,L \tag{5.5b}$$

where W = weight, lb (kg), d = internal diameter, in. (mm), t = wall thickness, in. (mm), L = shell length, ft (m).

The weight of one 2:1 ellipsoidal head is approximately:

Field units

$$W \approx 0.34td^2 + 1.9td. \tag{5.6}$$

The weight of a cone is

$$W = \frac{0.23td^2}{\sin\alpha}. \tag{5.7a}$$

SI units

$$W \approx 9.42 \times 10^{-6}td^2 + 1.34 \times 10^{-3}td, \tag{5.7b}$$

The weight of a cone is

$$W \approx 6.37 \times 10^{-6}\frac{td^2}{\sin\alpha},$$

where α = one-half the cone apex angle.

The weight of nozzles and internals can be estimated at 5–10% of the sum of the shell and head weights. As a first approximation, the weight of a skirt can be estimated as the same thickness as the shell (neglecting the corrosion allowance) with a length given by Equation (5.8) for an ellipsoidal head and Equation (5.9) for a conical head. For very tall vessels the skirt will have to be checked to assure it is sufficient to support both the weight of the vessel and its appentorances and the overturning moment generated by wind forces.

Field units

$$L = \frac{0.25d}{12} + 2, \tag{5.8a}$$

$$L = \frac{0.5d}{12\tan\alpha} + 2. \tag{5.8b}$$

SI units

$$L = 2.5 \times 10^{-4}d + 0.61, \tag{5.9a}$$

$$L = 2.54 \times 10^{-4}\frac{d}{\tan\alpha} + 0.61, \tag{5.9b}$$

where L = skirt length in ft (m).

The weight of pedestals for a horizontal vessel can be estimated as 10% of the total weight of the vessel.

5.5 Specification and Design of Pressure Vessels

5.5.1 Pressure Vessel Specifications

Some companies summarize their pressure vessel requirements on a pressure vessel design information sheet such as the one shown in Figure 5.5. Some companies have a detailed general specification for the construction of pressure vessels, which defines the overall quality of fabrication required and addresses specific items such as

- Code compliance
- Design conditions and materials
- Design details
  - Vessel design and tolerances
  - Vessel connections (nozzle schedules)
  - Vessel internals
  - Ladders, cages, platforms, and stairs
  - Vessel supports and lifting lugs
  - Insulation supports
  - Shop drawings
- Fabrication
  - General
  - Welding

- Painting
- Inspection and testing
- Identification stamping
- Drawings, final reports, and data sheets
- Preparation for shipment

A copy of this specification is normally attached to a bid request form, which includes a pressure vessel specification sheet such as the one shown in Figure 5.6. This sheet contains schematic vessel drawings and pertinent specifications and thus defines the vessel in enough detail so the manufacturer can quote a price and so the operator can be sure that all quotes represent comparable quality. The vessel connections (nozzle schedules) are developed from mechanical flow diagrams. It is not necessary for the bidder to know the location of the nozzles to submit a quote or even to order material.

5.5.2 Shop Drawings

Before the vessel fabrication can proceed, the fabricator will develop complete drawings and have these drawings approved by the representative of the engineering firm and/or the operating company. These drawings are called shop drawings. They will show detailed vessel design and fabrication/welding, nozzle schedules and locations, details of vessel internals, and other accessories. Examples are shown in Figures 5.7–5.15. Some typical details are discussed next.

5.5.3 Nozzles

Nozzles should be sized according to pipe sizing criteria, such as those provided in API RP 14E. The outlet nozzle is generally the same size as the inlet nozzle. To prevent baffle destruction due to impingement, the entering fluid velocity is to be limited as

Field units

$$V_{\text{in}} \leq (3500/\rho_f)^{1/2} \tag{5.10a}$$

SI units

$$V_{\text{in}} \leq (5217.7/\rho_f)^{1/2}$$

where V_{in} = maximum inlet nozzle fluid velocity, ft/s (m/s), p_f = density of the entering fluid, lb/ft^3 (kg/m^3).

If an interior centrifugal (cyclone) separator is used, the inlet nozzle size should be the same size as the pipe. If the internal design requires the smallest inlet and exit pressure losses possible, the nozzle size should be increased.

SEPARATOR DESIGN INFORMATION

1. Operating Conditions:
   A. Liquid Volumes
      1. Oil/Condensate: ___________ Barrel/Day Gravity: _____________ °API
      2. Water: ___________ Barrels/Day Sp. Gr.:_____________ (Water = 1.0)
   B. Oil/Condensate Characteristics
      1. Roaming: Nil ________ Moderate ________ Severe _________
      2. Paraffin Problem: No _____Yes_____
      3. Slug Flow: No _____Yes: ____ (If Yes, give details such as maximum liquid rate, slug volume, etc.)____
      ____
   C. Gas: __________MMscfd Sp. Gr.: __________ (Air - 1.0)
   D. Operating Temperature (°F):_______Max_______ Min
   E. Operating Pressure ipsig):_______ Max_______ Min
   F. H.S Content: _______Mole% CO_2. Content: _______. Mole%
   G. Geographical Location: ____

II Design Requirements:
   A. Type:——————Vertical________ Horizontal_________ Spherical.
      Manufacturer's Recommendation: _____________________________________
      _______Two-Phase _______Three-Phase
   B. Design Pressure: _________ psig at Temperature _________°F
   C. Type Mist Extractor: (Specify) _____________________________________
      ____
   D. Corrosion Allowance: _________(inches)
   E. Corrosion Allowance for Non-Pressure Internal Parts: ________(inches)
   F. NACE MR-01-75 Required: No_____ Yes_____
   G. Special Stress Relieving: No_____ Yes_____
      Specify if Yes: ____
      ____
   H. API RP 14C Safety Systems Required: No_____ Yes_____

III. Coatings:
   A. External: Mfgr. Std._______ Other_______
      Specify if Other: ____
   B. Internal: (Specify) ____
   C. Cathodic Protection: (Specify)_____________________________________
      ____

IV. Special Instructions:
   A. Radiographic Inspection: ASME Code ________ Other ________
      Specify if Other: ____
      ____
      ____
      ____
   B. Hydrostatic Test Pressure: ASME Code _________Other ________
      Specify if Other: ____
   C. Hardness Testing Requirements: (Specify) _________________________
   D. Lifting Lugs: (Specify) ____
   E. Skid Mounting: (Specify) ____
   F. Welding Requirement: ASME Code ________ Other ___________________
      Specify if Other: ____
   G. Sand Removal System: (Specify) ___________________________________
   H. Other: ____
      ____
      ____

FIGURE 5.5. Example of separator design information sheet.

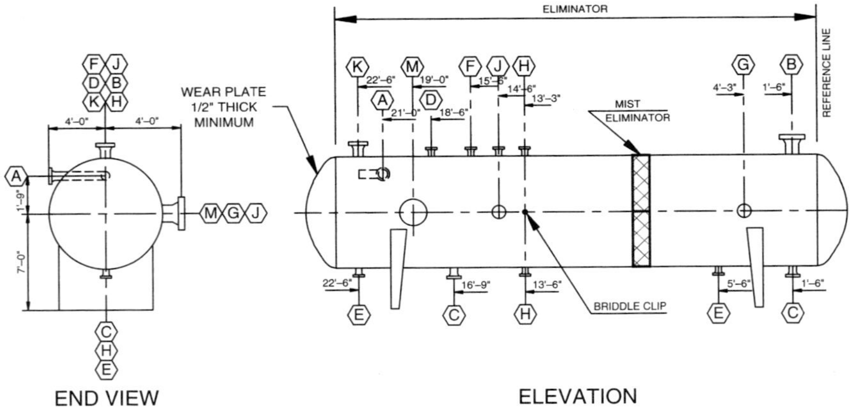

ITEM. *GAS SCRUBBER* ITEM NO. *MBD-1020*
NO. REQ'D. JOB NO.
PURCHASE ORDER NO. DATE.

DESIGN AND FABRICATION DATA

CONSTRUCTION TO BE IN ACCORDANCE WITH THE LATEST EDITION OF THE ASME CODE & ADDENDA. *SECTION VIII, DIVISION 2*
CODE SYMBOL. REQUIRED/NOT REQUIRED
DESIGN PRESSURE. 1800 PSIG. AT. –20/100 °F
OPERATING PRESSURE. 1000–1250 PSIG. AT. 60 °F
STRESS RELIEVE. YES/NO/PER CODE RADIOGRAPH. NP/SPOT/100%
JOINT E F F - SHELL. 1.0 HEADS. 1.0
CORROSION ALLOWANCE ALLOWANCE - SHELL 0.125" HEADS. 0.125°
MATERIAL: SHELL. HEADS. *SA - 516 - 70 2:1 ELLPT*
FLANGES. *SA - 516 - 70N* PIPE. *SA - 106 - B*
STUDS. *NOTE 3* NUTS. *SA - 194 - 2H*
GASKETS. *SA - 193 - B7*
SADDLES. *SOFT IRON TYPE R.I.D. MARK "D" CADMIUM PLATED*
HINGES. *YES* LUGS. *YES (2)*
INSULATION THICKNESS. *YES* DAVITS REQUIRED FOR MANHOLES. *NO*
LADDER CLIPS. *NONE* INSULATION RINGS. *NONE*
POINT PER SPEC. *NO* PLATFORM CLIPS. *REQUIRED*

DEMISTER. *NOTE 2*

NOTES:
1. DESIGN, FABRICATIONS, TESTING AND DOCUMENTATION SHALL BE IN ACCORDANCE WITH PARAGON SPECIFICATION
2. THE VANE TYPE MIST ELIMINATOR SHALL BE MANUFACTURED BY ACS INDUSTRIES, INC. (OR APPROVED EQUAL) AND SHALL REMOVE 99% OF ALL DROPLETS 10 MICRONS AND LARGER
3. WELD NECK FLANGES SHALL BE ASTM SA105 INTEGRALLY REINFORCED LONG WELD NECKS ARE ACCEPTABLE

NOZZLE SCHEDULE

MK	NO	SIZE	RATING	TYPE	SERVICE	PROJ.
A	1	12"	900#	RTJ	GAS/CONDENSATE INLET	–
B	1	12"	900#	RTJ	GAS OUTLET	12"
C	2	6"	900#	RTJ	CONDENSATE OUTLET	10"
D	1	2"	900#	RTJ	RELIEF/BLOWDOWN	8'
E	2	3"	900#	RTJ	DRAIN	8'
F	1	2"	900#	RTJ	PRESSURE CONNECTION	8'
G	1	2"	900#	RTJ	TEMPERATURE CONNECTION	8'
H	2	3"	900#	RTJ	LEVEL BRIDDLE	8'
J	2	2"	900#	RTJ	LEVEL BRIDDLE	8'
K	1	8"	900#	RTJ	INSPECTION W/BLIND	10"
M	1	18"	900#	RTJ	MANWAY 18" I.D.	–

PROCESS CONDITIONS

GAS FLOW RATE 200 MMSCFD
GAS SPECIFIC GRAVITY: 0.67 (AIR = 1.0)
HYDROCARBON LIQUID FLOW RATE: 2.5 BBL/MMSCF NORMAL
HYDROCARBON LIQUID SPECIFIC GRAVITY: 0.56 @ OPERATING CONDITIONS (WATER = 1.0)
OPERATING PRESSURE: 1000 PSIG MINIMUM, 1250 MAXIMUM
OPERATING TEMPERATURE: 55°F MINIMUM, 70°F MAXIMUM

NOTES
1. INTERNAL INLET PIPING SHALL BE DESIGNED TO WITHSTAND LIQUID SLUGS ARRIVING AT VELOCITIES AS HIGH AS 45 FT/SEC.
2. VESSEL ORDINARILY OPERATES EMPTY, BUT LIQUID LEVEL DURING SLUGGING CAN BE AS HIGH AS 42° ABOVE OUTSIDE BOTTOM OF VESSEL

NO.	REVISION	DATE	DRAWN	CHECK	APP'D

ISSUED FOR: CLIENT APPROVAL / BODING / CONSTRUCTION
ENGINEER. DATE. DRAWN. DATE. CHECKED. DATE. APPROVED. DATE. SCALE. SHEET. OF JOB NO. CLIENT. CLIENT JOB NO.

PARAGON ENGINEERING SERVICES HOUSTON, TEXAS

PARAGON ENGINEERING SERVICES
MBD - 1020
LP PRODUCTION SEPARATOR
DRAWING NO. MBD - 1020 REV.

FIGURE 5.6. Example of pressure vessel specification sheet.

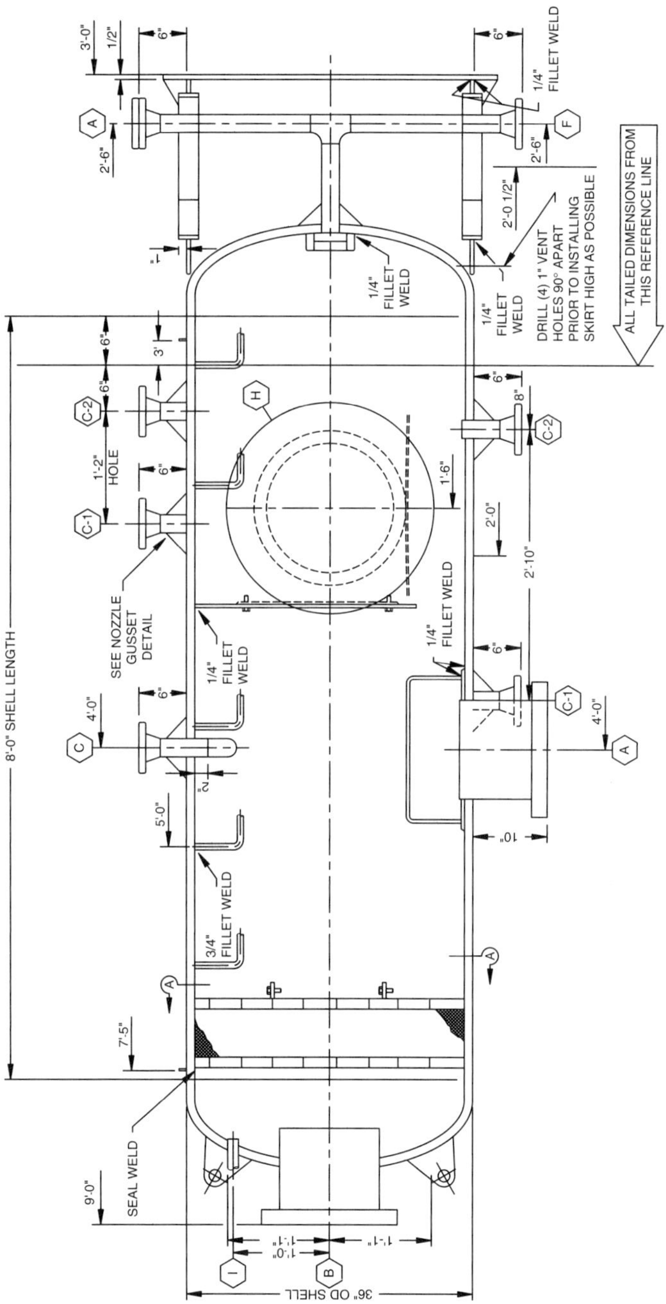

FIGURE 5.7. Example of pressure vessel shop drawing.

Outside projection, inches using welding neck flange

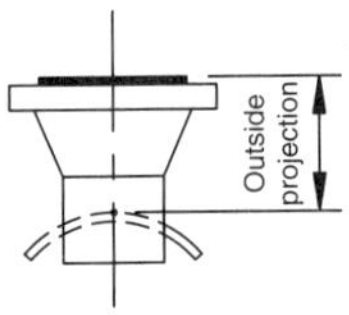

Nom. pipe size	Pressure rating of flange LB					
	150	300	600	900	1500	2500
2	6	6	6	8	8	8
3	6	6	8	8	8	10
4	6	8	8	8	8	12
6	8	8	8	10	10	14
8	8	8	10	10	12	16
10	8	8	10	12	14	20
12	8	8	10	12	16	22
14	8	10	10	14	16	
16	8	10	10	14	16	
18	10	10	12	14	18	
20	10	10	12	14	18	
24	10	10	12	14	20	

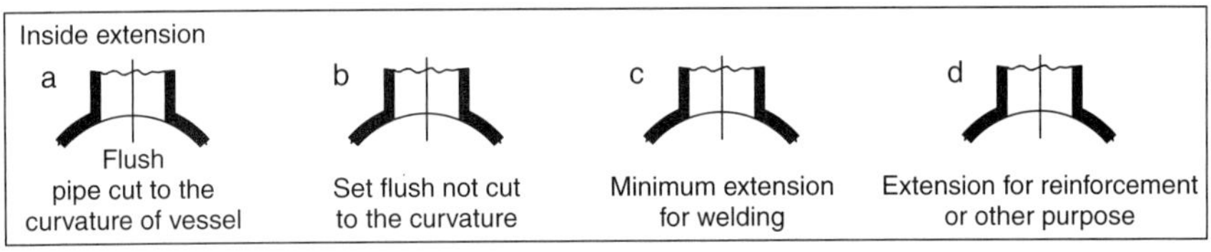

FIGURE 5.8. Nozzel projections. (Reprinted with permission from Pressure Vessel Handbook, Publishing, Inc., Tulsa.)

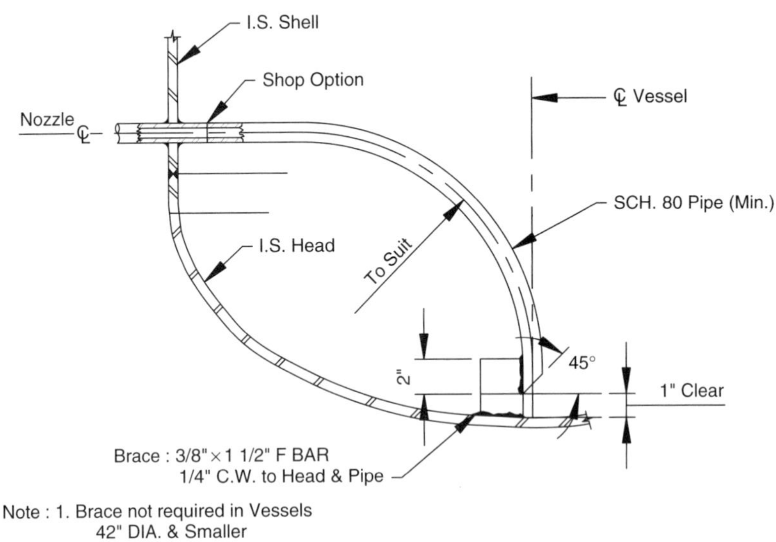

FIGURE 5.9. Siphon drain.

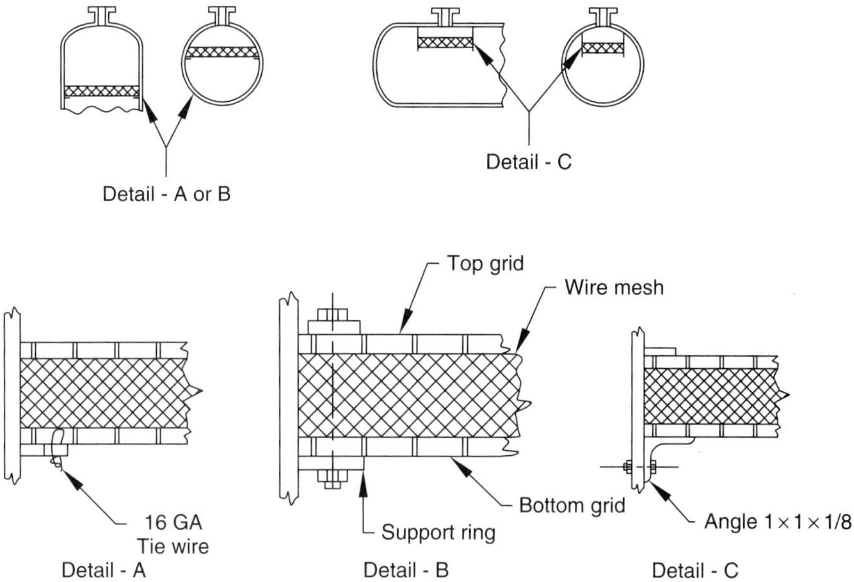

FIGURE 5.10. Example of supports for mist extractors. (Reprinted with permission from Pressure Vessel Handbook, Publishing, Inc., Tulsa)

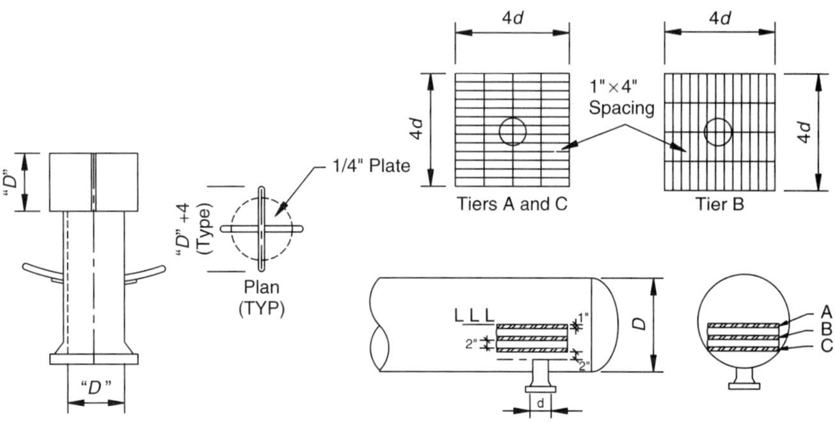

FIGURE 5.11. Examples of Vortex Breaker Details.

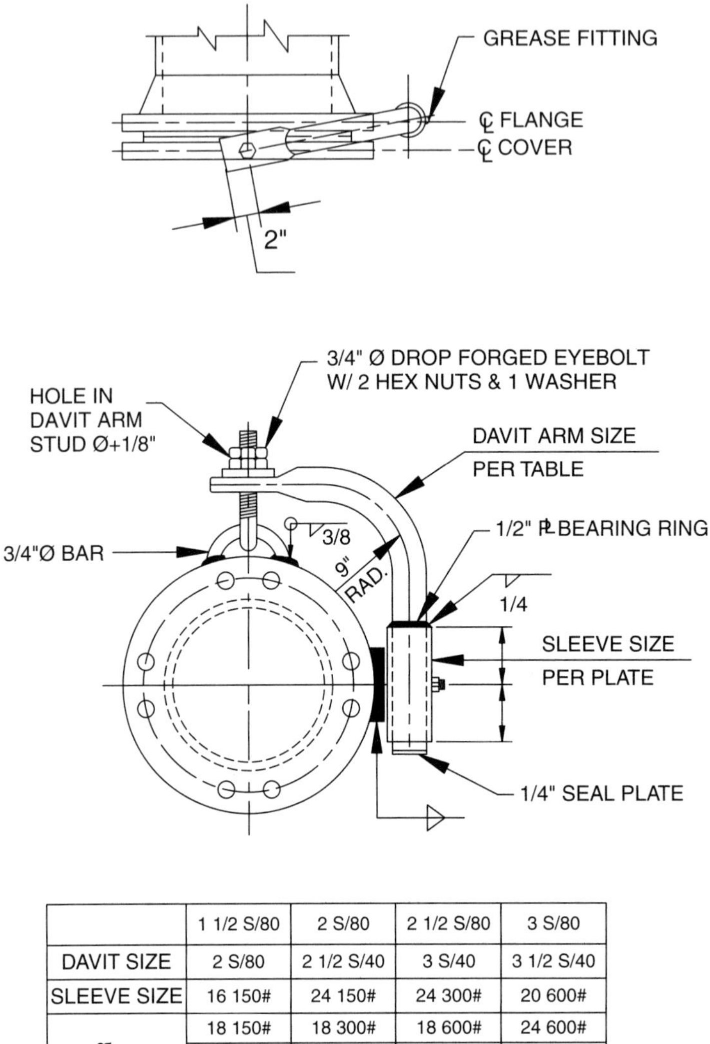

	1 1/2 S/80	2 S/80	2 1/2 S/80	3 S/80
DAVIT SIZE	2 S/80	2 1/2 S/40	3 S/40	3 1/2 S/40
SLEEVE SIZE	16 150#	24 150#	24 300#	20 600#
MANWAY COVER SIZE & RATING	18 150#	18 300#	18 600#	24 600#
	20 150#	20 300#	20 600#	16 900#
	16 300#			18 900#
				20 900#
	ON ANY COVER NOT EXCEEDING			
	325# IN WEIGHT	525# IN WEIGHT	850# IN WEIGHT	1200# IN WEIGHT

FIGURE 5.12. Examples of horizontal manway cover davit and sleeve detail.

BASE PLATE SCHEDULE		
ANGLE LEG SIZE	"A"	"B"
6" × 6"	8"	3 3/8"
5" × 5"	7"	2 7/8"
4" × 4"	6"	2"
3" × 3"	5"	1 3/4"
2 1/2" × 2 1/2"	4"	1 1/2"

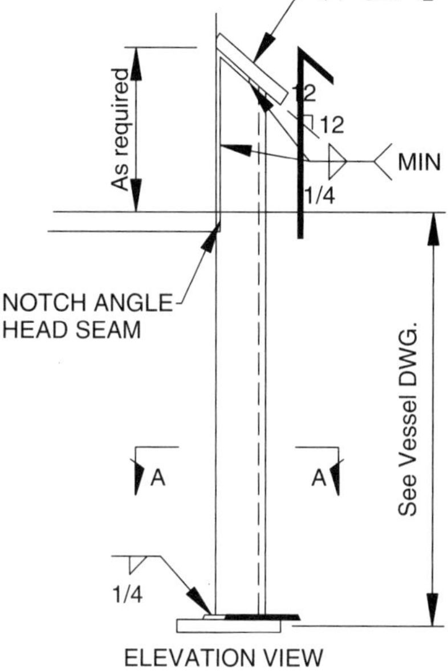

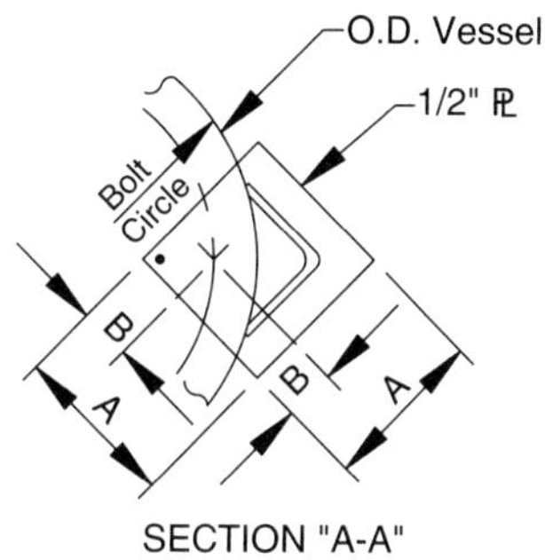

FIGURE 5.13. Angle support legs.

5.5.4 Vortex Breaker

As liquid flows out of the exit nozzle, it will swirl and create a vortex. Vortexing would carry the gas out with the liquid. Therefore, all liquid outlet nozzles should be equipped with a vortex breaker. Figure 5.11 shows several vortex breaker designs. Additional designs can be found in the *Pressure Vessel Handbook.* Most designs depend on baffles around or above the outlet to prevent swirling.

5.5.5 Manways

Manways are large openings that allow personnel access to the vessel internals for their maintenance and/or replacement. Vessels 36 in. and larger should have a minimum of one 18-in. manway. Vessels 30 in. and smaller should have two 4-in. flanged inspection openings. Manway cover davits should be provided for 12 in. and larger manways for safe and easy opening and closing of the cover. Figure 5.12 shows an example of a manway cover davit and sleeve details.

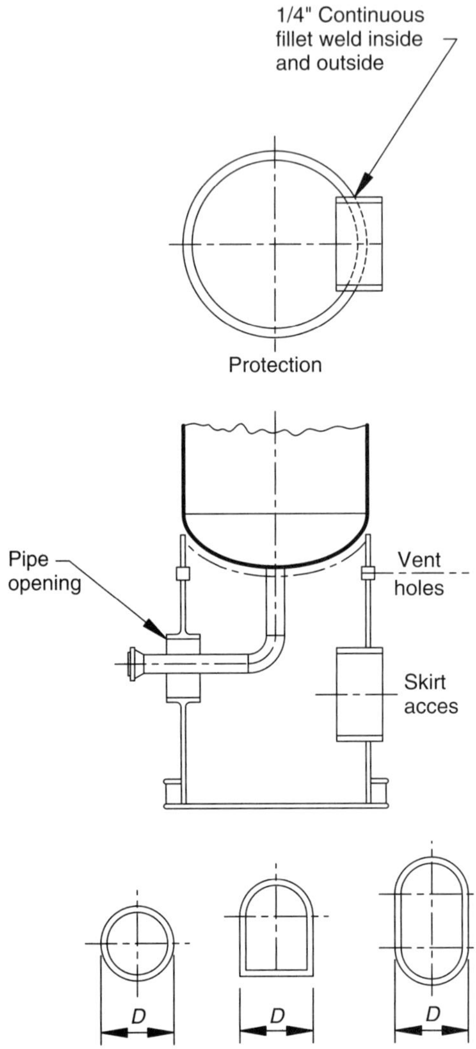

FIGURE 5.14. Skirt openings. (Reprinted with permission from Pressure Vessel Handbook, Publishing, Inc., Tulsa.)

5.5.6 Vessel Supports

Small vertical vessels may be supported by angle support legs, as shown in Figure 5.13. Larger vertical vessels are generally supported by a skirt support, as shown in Figure 5.14. At least two vent holes, 180× apart, should be provided at the uppermost location in the skirt to prevent the accumulation of gas, which may create explosive conditions. Horizontal vessels are generally supported by a pair of saddle-type supports.

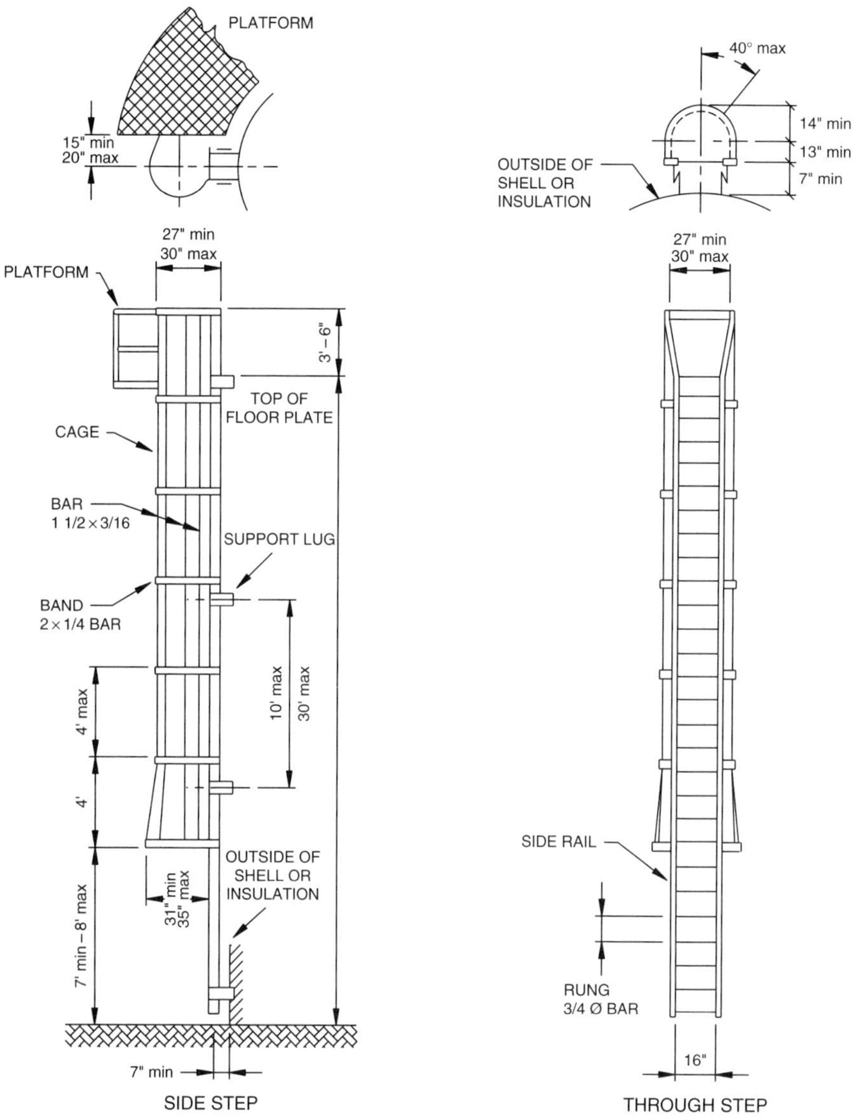

FIGURE 5.15. Ladders. (Reprinted with permission from Pressure Vessel Handbook, Publishing, Inc., Tulsa.)

5.5.7 Ladder and Platform

A ladder and platform should be provided if operators are required to climb up to the top of the vessel regularly. An example is shown in Figure 5.15.

WEIGHT OF SHELLS & HEADS										
	WALL THICKNESS									
DIAM.	1-3/8"					1-7/16"				
VESSEL	SHELL		HEAD			SHELL		HEAD		
	I.S	O.S	ELIP	F&D	HEMIS	I.S.	O.S.	ELLIP	F&D	HEMIS
12	146	156	142	119	135	206	162	151	126	143
14	225	185	169	148	178	237	193	180	155	188
16	255	215	206	176	228	267	223	220	184	240
18	284	244	239	206	283	289	254	255	220	298
20	313	273	285	239	345	329	285	303	253	363
22	343	303	322	275	412	360	316	342	292	434
24	372	332	364	323	486	390	346	286	337	511
26	402	362	408	364	566	421	377	432	380	594
28	431	391	466	408	651	452	408	493	426	684
30	460	421	527	454	743	482	438	558	481	780
32	490	450	593	502	841	513	469	627	532	882
34	519	479	662	553	945	544	500	699	585	991
36	548	508	734	620	1054	575	531	775	648	1106
38	578	538	812	676	1170	605	561	857	707	1228
40	607	567	892	812	1293	636	592	941	768	1355

FIGURE 5.16. Weight of shells and heads. (Reprinted with permission from Pressure Vessel Handbook, Publishing, Inc., Tulsa.)

5.6 Pressure Relief Devices

All pressure vessels should be equipped with one or more pressure safety valves (PSVs) to prevent overpressure. This is a requirement of both the ASME code and API RP 14C. The PSV should be located upstream of the mist extractor. If the PSV is located downstream of the mist extractor, an overpressure situation could occur when the mist extractor becomes plugged, isolating the PSV from the high pressure, or the mist extractor could be damaged when the relief valve opens. Rupture discs are sometimes used as a backup relief device for the PSV. The disc is designed to break when the internal pressure exceeds the set point. Unlike the PSV, which is self-closing, the rupture disc must be replaced if it has been activated.

5.7 Corrosion Protection

Pressure vessels handling salt water and fluids containing significant amounts of H_2S and CO_2 require corrosion protection. Common corrosion protection methods include internal coatings with synthetic polymeric materials and galvanic (sacrificial) anodes. All pressure vessels that handle corrosive fluids should be monitored periodically. Ultrasonic surveys can locate discontinuities in the metal structure, which will indicate corrosion damages.

Example 5.1. Determining the weight of an FWKO vessel (field units). Determine the weight for the following free-water knockout vessel. It is butt weld fabricated with spot x-ray and to be built to the ASME code Section VIII, Division 1, 1998 edition. A conical head (bottom of the vessel) is desired for ease in sand removal. Compare this weight to that of a vessel without the conical section and that to a vessel with a 0.25-in. plate internal cone.

Design pressure = 125 psig,
Maximum operating temperature = 200 °F,
Corrosion allowance = 0.25 in.,
Material = SA516 Grade 70,
Diameter = 10 ft,
Seam-to-seam length above the cone = 12 ft,
Cone apex angle = 60°.

Solution:

Case I—Cone Bottom

(a) Shell:

$$t = \frac{Pr}{SE - 0.6P},$$

$$S = 17,500 \text{ psi}, \qquad \text{(Table 5.3)}$$

$$E = 0.85, \qquad \text{(Table 5.4)}$$

$$t = \frac{(125)(60)}{(17,500)(0.85) - (0.6)(125)} = 0.507 \text{ in.},$$

Required thickness = 0.507 + 0.250 = 0.757 in., use 13/16-in. plate (0.8125 in.)

$$W = 11dtL = (11)(120)(0.8125)(12) = 12,870 \text{ lb}.$$

(b) Head (ellipsoidal 2:1):

$$t = \frac{(125)(120)}{(2)(17,500)(0.85) - (0.2)(125)} = 0.505 \text{ in.},$$

Required thickness = 0.505 + 0.250 = 0.755 in., use 13/16-in. plate (0.8125 in.)

$$W \approx 0.34td^2 + 1.9td,$$

$$W = (0.34)(0.8125)(120)^2 + (1.9)(0.8125)(120) = 4163 \text{ lb}.$$

(c) Cone:

$$t = \frac{Pd}{2\cos\alpha(SE - 0.6P)},$$

$$t = \frac{(125)(120)}{(2\cos 30)(17,500 \times 0.85 - 0.6 \times 125)} = 0.585 \text{ in.},$$

Required thickness = 0.585 + 0.250 = 0.835 in., use 7/8-in. plate (0.875 in.)

$$W = \frac{(0.23)(0.875)(120)^2}{\sin 30} = 5796 \text{ lb.}$$

(d) Skirt:

$$\text{Height} = \frac{5}{\tan 30} = 8.66 \text{ ft},$$

Allow 2 ft for access, Height = 11 ft (The shell wall thickness, neglecting corrosion allowance, is ~0.5 in. Assume 0.5-in. plate), $W = (11)(120)(0.5)(11) = 79{,}860$

(e) Summary:

Shell	12,870
Skirt	7260
Subtotal	30,089
Misc.	5000
Total	35,089 lb

Case II—2:1 Ellipsoidal Head

(a) Skirt:

$$L = \frac{0.25\,d}{12} + 2$$

$$= \frac{(0.25)(120)}{12} + 2$$

$$= 4.50 \text{ ft},$$

$$W = (11)(120)(0.5)(4.5)$$

$$= 2,970 \text{ lb.}$$

(b) Summary:

Shell	12,870
Head-1	4,163
Head-2	4,163
Skirt	2,970
Subtotal	24,166
Misc.	5000
Total	29,166 lb

Case III—Internal Cone

(a) Internal cone:

$$W = \frac{(0.23)(0.25)(120)^2}{\sin 30}$$
$$= 1656 \text{ ft.}$$

(b) Shell:

$$\text{Height of cone} = \frac{(10/2)}{\tan 30} = 8.7 \text{ ft},$$

$$\text{Length of shell} = 12 + 8.7 = 20.7 \text{ ft},$$

$$\text{Weight of shell} = (11)(120)(0.8125)(20.7)$$
$$= 22,200 \text{ lb.}$$

(c) Summary:

Shell	22,200
Head-1	4,163
Head-2	4,163
Skirt	2,970
Cone	1,656
Subtotal	35,152
Misc.	5000
Total	40,152 lb

Reference

1. Bednar, H. H., *Pressure Vessel Design Handbook*, Van Nostrand Reinhold, 2004.

Glossary of Terms

Acid gas H_2S and/or CO_2 contained in or extracted from a natural gas.

Accumulator A vessel used to collect and store liquids.

API gravity An arbitrary scale expressing the relative density of liquid petroleum products. The measuring scale is calibrated in degrees API (°API) and is calculated by the following formula:

$$°API = \frac{141.5}{SG\ 60°F/60°F} - 131.5.$$

Artificial lift Mechanical means of raising a crude oil in a well to the surface, including sucker-rod pump, hydraulic pump, gas lift, and electrical submersible pump.

Atmospheric pressure The pressure exerted on the earth by the earth's atmosphere. A pressure of 760 mmHg, 29.92 in. of mercury, or 14.696 psia is used as a standard for some measurements. The various state regulatory bodies have set other standards for use in measuring the legal volume of natural gas that is sold or processed. Atmospheric pressure may also refer to the absolute ambient pressure at any given location.

Bad oil Crude with a pipeline spec. BS&W content in excess of Boiling point.

Boiling range Range of boiling point temperatures used to characterize a cut.

Bubble point The temperature at a given pressure or the pressure at a given temperature at the instant the first bubble of gas is formed in a given liquid.

Cannula A large-bore hypodermic needle attached to a syringe; used to remove samples from liquid layers.

Chromatography A technique for sample analysis where individual components of a batch sample, carried by an inert gas stream, are selectively sorbed and disrobed on a sorbent column at different rates in relation to equilibrium coefficients. Separated components are quantitatively detected as they leave the sorbent column.

Clean crude Crude oil containing no BS&W.

Collector pipe Perforated or slotted pipe used to remove treated oil as uniformly as possible at top of coalescing section.

Compressibility factor A factor usually expressed as Z, which gives the ratio of the actual volume of gas at a given temperature and pressure to the volume of gas when calculated by the ideal gas law without any consideration of the compressibility factor.

Conditioning See "processing."

Connate water Formation water held in the pores by capillary action; water originally contained in sedimentary rocks at the time of deposition.

Continuous phase See "emulsion."

Control valve Valve used to control flow rate of a fluid entering or leaving a process component.

Convergence pressure The pressure at given temperature for a hydrocarbon system of fixed composition at which the vapor–liquid equilibria values of the various components in the system become unity. The convergence pressure is used to adjust the vapor-liquid system under consideration.

Cricondenbar The highest pressure at which vapor and liquid phases can be identified in a multi-component system.

Cricondentherm The highest temperature at which vapor and liquid phases can be identified in a multi-component system.

Critical pressure The pressure necessary to condense a vapor at its critical temperature.

Critical temperature The highest temperature at which a pure element or compound can exist as a liquid. Above this temperature, the fluid is a gas and cannot be liquefied regardless of the pressure applied.

Crude oil Unrefined liquid petroleum.

Cubic equation Equation of state with three constants.

Custody transfer Transfer of ownership of oil or gas streams, usually at some arbitrary location in the field.

Cut A petroleum fraction containing numerous individual compounds that is characterized by average properties such as boiling point range, API, SG, and so on.

Cyclone A cone-shaped separator that uses centrifugal force to separate two immiscible phases.

Dehydration The act or process of removing water from gases or liquids.

Demulsifier Demulsifiers or demulsifying chemicals are a mixture of chemicals used to break the emulsion by destroying or weakening the stabilizing film around the dispersed drops.

Dense phase Fluid existing above both the cricondenbar pressure and the critical temperature.

Desalting The act or process of removing salts from crude oils.

Desulfurization The process by which sulfur and sulfur compounds are removed from gases or liquid hydrocarbon mixtures.

Dew point The temperature at any given pressure or pressure at a given temperature at which liquid initially condenses from a gas or vapor. It is specifically applied to the temperature at which the water vapor starts to condense from a gas mixture (water dew point) or at which hydrocarbon starts to condense (hydrocarbon dew point).

Direct heater A heater in which fire-tube contacts the process fluid directly.

Dispersed phase See "emulsion."

Drive Pressure tending to cause an oil in reservoir to flow through the rock pores to the well bore and upwards through the tubing to the surface; common types of drive are free gas cap, dissolved gas, water, and gravity.

Dry gas (1) Gas containing little or no hydrocarbons commercially recoverable as liquid product. Gas in this definition preferably should be called "lean gas." (2) Gas whose water content has been reduce by a dehydration process (rare usage).

Dual emulsion An emulsion in which the continuous phase is oil and the dispersed phase is an oil-in-water emulsion.

Electrodes or grid Plates or rods used to establish the electric field in electrostatic treaters.

Electrostatic Treater using electrostatic fields in the oil treater coalescing area.

Emulsified water Water that will not separate readily from a water-in-crude emulsion.

Emulsifier In addition to oil and water, a third substance—called an emulsifier or emulsifying agent—must be present for a stable emulsion to be produced. These emulsifiers usually exist as a film on the surface of the dispersed drops.

Emulsion A combination of two immiscible liquids. One liquid is broken up into droplets and is known as the discontinuous, dispersed, or internal phase. The other liquid that surrounds the drops is the continuous or external phase.

Equation of state An equation relating the pressure, temperature, and specific volume of a fluid.

Error Set-point value—process output.

Excelsior Fibrous material used to separate water from oil in a heater-treater.

External phase See "emulsion."

Flash point The lowest temperature at which vapor from a hydrocarbon liquid will ignite.

Free water Water that separates readily (in <5 min) from a produced crude oil.

Gain Ratio of controller output to error.

Gas anchor A short section of tubing that extends down from an insert sucker-rod pump and is used to separate gas from oil before it enters the pump to prevent gas locking.

Gas-condensate field A petroleum field or reservoir in which the hydrocarbons in the formation exist in a vapor state under high temperature. A lowering of the temperature causes a condensation of the heavier hydrocarbons, which will then not be produced with the gas.

Gas constant A constant number, which mathematically is the product of the total volume and the total pressure, divided by the absolute temperature for one mole of any ideal gas or mixture of ideal gases at any temperature. $PV/T=R$.

Gathering lines The network of pipelines that carry gas/oil from the wells to the processing plant or other separation equipment.

Gauging Measurement of oil in a storage tank.

Grasshopper Vertical pipe arrangement on the outside of an atmospheric crude oil tank that controls internal water–oil interfacial level by manipulation of its height.

Gun barrel Settling tank or wash tank, with built-in gas boot.

Handling See "processing."

Hay See "excelsior."

Head Pressure due to a height of fluid.

Heater-treater A vessel used to dehydrate crude oil that uses chemicals, settling, and heat.

Heating baffle A baffle that surrounds the fire-tubes and is hood or shroud designed to minimize heating of free water in a heater-treater.

Heating value The amount of heat developed by the complete combustion of a unit quantity of a material.

Heave Vertical motion of a ship or floating platform.

Hexane (or Heptanes) plus The portion of a hydrocarbon fluid mixture or the last component of a hydrocarbon analysis that contains the hexanes (or heptanes) and all hydrocarbons heavier than the hexanes (or heptanes).

Hydrate A solid material resulting from the combination of hydrocarbon with water under pressure.

Indirect heater A heater in which the fire-tube heats a liquid that, in turn, heats the process fluid.

Injection of gas Putting gas into the formation by force (pressure).

Innage Crude oil contained in a tank between the tank bottom and the oil surface; as contrasted to outage (see "outage").

Interface Two uses: (1) the surface area of the drops in an emulsion; (2) the area between two separated phases in a vessel.

Interface pad A layer of solid accumulated at the interface between relatively pure water and oil layers.

Internal phase See "emulsion."

Interphase drain A perforated pipe or other device used to remove the solid phase accumulated at the oil–water interface in a treater.

Inverse emulsion See "reverse emulsion."

Joule–Thomson The change in gas temperature that occurs when the gas is expanded at constant enthalpy from a higher pressure to a lower pressure. The effect for most gases at normal pressure, except hydrogen and helium, is a cooling of the gas.

***K* value** Ratio of mole fraction of a component in vapor to that in liquid.

Knockout Separator that removes (1) free water from crude oil or (2) total liquids from a gas stream.

Knockout drops A demulsifier used to separate BS&W from a crude oil emulsion sample; allows determination of BS&W.

Lean gas (1) The residue gas remaining after recovery of natural gas liquids in a gas processing plant. (2) Unprocessed gas containing little or no recoverable natural gas liquids.

Light ends The low-boiling, easily evaporated components of a hydrocarbon liquid.

Loose emulsion An unstable or easily broken emulsion.

Manifold A pipe with one or more inlets and two or more outlets, or vice versa.

Mercaptan A compound sometimes found in gas and gas liquids which must be reduced by removal or conversion to conform to specification. Any of a series of compounds of the alcohol and phenols, but containing sulfur in place of oxygen. (*R* represents an alkyl group or radical.)

Molecular sieve A synthetic zealot (essentially silica–alumina) used in adsorption processes.

Natural gas Gaseous petroleum.

Offset Set-point—process output after control action.

Oil-field Surface area overlying an oil reservoir.

Oil-in-water An emulsion consisting of oil drops dispersed in (o/w) emulsion a continuous water phase.

Outage Space in a tank between the oil surface and the top of the tank; also called "ullage."

Overdosing Adding excess or too much demulsifier.

Plate-fin exchangers Heat exchangers, which use thin sheets of metal to separate the hot and cold fluids instead of tubes.

Pentane-plus A hydrocarbon mixture consisting mostly of normal pentane (C_5H_{12}) and heavier components extracted from natural gas.

Petroleum Hydrocarbons (gas and oil) obtained from underground reservoirs.

Pigging A procedure of forcing a solid object through a pipeline for cleaning purposes.

Pipeline oil A crude oil that meets all pipeline specs such as API, S content, pour point, S&W content, RVP, etc.

Pitch Angular motion of a ship or floating platform.

Pressure maintenance Injection of gas into a formation to keep up the pressure.

Processing All unit operations performed on wellhead fluids in the field.

Produced water Water produced with crude oil or gas. It is usually classified as entrained or free. Entrained or emulsified water does not settle out readily. Free water settles within 5 min.

Proportional band 100 Controller Gain

Prover Device used to calibrate a flow meter.

Raw gas Unprocessed gas or the inlet gas to a plant.

Raw mix liquids A mixture of natural gas liquid prior to fractionation. Also called "raw make."

Recompressor A compressor used from some particular service, such as compressing residue gas; implies restoring of pressure level of a stream that has been subjected to pressure reduction.

Regular emulsion A water-in-oil (w/o) emulsion.

Relief system The system for temporarily releasing excess fluid, usually gas, to avoid a pressure in excess of the design pressure for the particular equipment.

Reservoir Subsurface, permeable rocks body containing crude oil and/or natural gas.

Retrograde condensate (vaporization) Condensate or vaporization that is reverse of usual behavior. Condensation caused by a decrease in pressure or increase in temperature. Vaporization caused by an increase in pressure or decrease in temperature. Can only occur in mixtures.

Reverse emulsion An oil-in-water (o/w) emulsion.

Roll Angular motion of a ship or a floating platform.

RVP (Reid vapor pressure) A vapor pressure for liquid products as determined by ASTM test procedure D-323. The Reid vapor pressure is reported as pound per square inch at 100 °F. The RVP is always less than the true vapor pressure at 100 °F.

Sales gas A gas that meets all specifications for sales.

Sand pans Inverted troughs or angle's baffles used to aid sand and sediment removal from treaters.

Scrubber A separator that removes small amounts of liquid from a gas stream.

Sensor Measuring instrument.

Separator Vessel used to split a multi-phase well stream into a gas stream and one or more liquid streams.

Separator gas Same as associated gas.

Shrinkage Reduction in volume of oil as gas is evolved from it.

Solution gas Gas that is dissolved in crude oil, either in a reservoir or in the producing equipment.

Sour gas or oil A gas or oil containing H_2S or mercaptans above a specified concentration level.

Specific gravity The ratio of the mass of given volume of a substance to that of an equal volume of another substance used as standard. Unless otherwise stated, air is used as the standard for gases and water for liquids and the volumes measured at 60 °F and atmospheric pressure (15.56 °C and 101.325 kPa).

Spreaders Perforated pipes or channels used to inject emulsions as uniformly as possible throughout the treater's cross section.

Stabilization Removing volatile compound from a crude oil to reduce its bubble-point pressure (and its RVP).

Stabilizer A name for a fractionation system that stabilizes any liquid (i.e., reduces the vapor pressure so that the resulting liquid is less volatile).

Stable emulsions Require an active treatment for breaking or phase separation to occur.

Steam flooding EOR method for shallow, heavy oil deposits in which high-temperature steam is injected into the formation to make the oil more easily produced.

Stock-tank oil Oil remaining after stage-separation train or stabilization (i.e., after dissolved gas has been released).

Strapping Measuring and recording the dimension of a storage tank.

Sulfur A yellow, non-metallic chemical element. In its elemental state, called "free sulfur," it has a crystalline or amorphous form.

In many gases and oil streams, sulfur may be found in volatile sulfur compounds (i.e., hydrogen sulfide, sulfur oxides, mercaptans, carbonyl sulfide).

Surge Motion of a ship or floating platform; pressure pulse in a pipeline.

Surge factor Equipment is usually sized using the maximum flow rate expected during predicted life of facility. Generally, accepted practice is to add a surge factor (20–50%) to handle short-term fluctuations.

Sway Motion of a ship or floating platform.

Sweet This refers to the near or absolute absence of objectionable sulfur compounds in either gas or liquid as defined by given specification standard.

Sweetening Act or process of removing H_2S and other sulfur compounds.

Tight emulsion A very stable or hard-to-break emulsion.

Trap Gas–oil separator, usually horizontal.

Treating Removing undesirable components or properties from a fluid.

Vapor pressure The pressure exerted by a liquid when confined in a specified tank or test apparatus.

V/L ratio Vapor–liquid equilibrium ratio.

Water cut Volume % water in crude oil–water mixture.

Water-in-oil In vast majority of cases, crude oil emulsions consist of an emulsion of water drops dispersed in a continuous oil phase. Also called "regular" or "normal emulsion."

Water leg Piping system for removing water from a separator by overflowing an external or internal weir. Also called "grasshopper."

Wet gas Natural gas that yields hydrocarbon condensate (does not usually refer to water content). Also called "rich" gas.

Wetting Refers to adhesion or sticking of a liquid to a solid surface. If the solid surface (grain of reservoir rock, fines, etc.) is covered preferentially by oil, the surface is called "oil wetted." If water is preferentially attracted, the surface is "water wetted."

Yaw Angular motion of a ship or floating platform.

Common Abbreviations

ACT	Automatic custody transfer; see LACT
AG	Acid gas
AGA	American Gas Association
AIME	American Institute of Mining, Metallurgical, and Petroleum Engineers
AISI	American Iron & Steel Institute
ANSI	American National Standards Institute
API	American Petroleum Institute—National Trade Association of United States Petroleum Industry, a private standardizing and lobbying organization
ASME	American Society of Mechanical Engineers
ASTM	American Society for Testing and Materials
ATG	Automatic tank gauging system
atm	Atmosphere
bbl	Barrel (42 U.S. gallons). The oil industry standard for volumes of oil and its products; always reduced to 60 °F and vapor pressure of the liquid
BEP	Best efficiency point (for a centrifugal pump)
Bhp	Brake horsepower
BLM	Bureau of Land Management—U.S. government agency that regulates petroleum production onshore
blpd	Barrels of liquid per day
B_o	Formation volume factor
BOPD	Barrels of oil per day
BPD	Barrels per day
Brf	Barrels of reservoir fluid
BS&W	Basic sediment and water; water and other contaminants present in crude oil
BTEX	Benzene, toluene, ethyl benzene, and xylene
Bscf	Billions of standard cubic feet
bsto	Barrels of stock-tank oil
BTU	British thermal unit
BWPD	Barrels of water per day
C_1	Methane
C_2	Ethane
C_3	Propane
C_4's	Butanes
C_5's	Pentanes
C_6	Hexanes
C_6+	Hexanes and heavier
C_7	Heptanes
C_7+	Heptanes and heavier

C_8	Octanes
CAAA	Clean Air Act Amendments
CF	Characterization factor
cfm	Cubic feet per minute
CI	Controller input
CMA	Chemical Manufacturers Association
CMV	Corrected meter volume
CO	Controller output
cp	Centipoise
CV	Control valve
CW	Continuous-welded
API	API gravity
Degrees	
F Degrees	Fahrenheit
C Degrees	Celsius
DOE	Department of Energy
DOT	Department of Transportation
EBHAZOP	Experienced-based HAZOP
ECT	Environmental control technology
EOR	Enhanced oil recovery
EPA	Environmental Protection Agency
EODR	Electro optical distance ranging
EOS	Equation of state
ERW	Electric resistance welded
ft/sec	Feet per second
FERC	Federal Energy Regulatory Commission
FIA	Fire Insurance Association
FMA	Factory Mutual Association
FRP	Fiber-reinforced plastic
FVF	Formation volume factor
FWKO	Free-water knockout
gal	U.S. gallon
GHV	Gross heating value
GLC	Gas–liquid chromatography
GLR	Gas–liquid ratio, expressed as scf/bbl
GOM	Gulf of Mexico
GOR	Gas–oil ratio, combined gas released from stage separation of oil, expressed as scf/Bsto
GOSP	Gas–oil separation plant
gph	Gallons per hour
GPM	Gallons liquefiable hydrocarbons per 1000 scf of natural gas
gpm	Gallons per minute; describes liquid flow rate
GPSA	Gas Processors Supplier Association
gr	Grain (7000 gr=1 lb)

GSC	Gas–solid chromatography
HAZIN	Hazards identification
HAZOPS	Hazards Operability Study
HC	Hydrocarbon
HCL	Higher combustion limit
HHV	Higher heating value
HP	High pressure
hp	Horsepower
hp-h, hp-hr	Horsepower-hour
HTG	Hydrostatic tank gauging
H_2O	Water
H_2S	Hydrogen sulfide
i-C_4	Isobutane
i-C_5	Isopentane
ID	Inside diameter
ISA	Instrument Society of America
ISO	International Standards Organization
J–T	Joule–Thomson (constant enthalpy) expansion
kW	Kilowatts
kWh	Kilowatts-hour
LACT	Lease automatic custody transfer
LC	Level control
LCL	Lower combustion limit
LCV	Level control valve
lb	Pounds
lbmol	Pound mole
LED	Light emitting diode
LET	Lowest expected temperature
LHV	Lower heating value
LMTD	Log mean temperature difference
LNG	Liquefied natural gas; primarily C_1 with lesser amounts of C_2 and C_3
LP	Low pressure
LPG	Liquefied petroleum gas, C3-C4 mix
mA	Milliampere
MAWP	Maximum allowable working pressure
Mcf	Sloppy equivalent for Mscf
Mcfd	Thousand cubic feet per calendar day
MF	Meter factor
MIGAS	Ministry of Oil and Gas (Indonesia)
MMcf	Same as MMscf
MMcfd	Millions of standard cubic feet
MMscfd	MMscf per day
MMS	Minerals Management Service

MPT	Minimum pipeline temperature
Mscf	Thousand standard cubic feet
Mscfd	Mscf per day
MW	Molecular weight
N, N_2	Nitrogen
NACE	National Association of Corrosion Engineers
NBS	National Bureau of Standards, now NIST
n-C_4	Normal butane
n-C_5	Normal pentane
NFPA	National Fire Protection Association
NGL	Natural gas liquids; includes ethane, propane, butanes, pentanes, or mixture of these
NHV	Net heating value
NIST	National Institute for Standard and Technology, formerly NBS
NORM	Naturally occurring radioactive materials
NPDES	National Pollution Discharge Elimination System
NPS	National pipe standard
NPSH	Net positive suction head
NPSHA	Net positive suction head available
NPSHR	Net positive suction head required
OCS	Outer continental shelf
OD	Outside diameter
ORLM	Optical reference line method
OSHA	Occupational Safety and Health Administration
OTM	Optical triangulation method
PCV	Pressure control valve
PD	Positive displacement (e.g., a PD pump)
PE	Polyethylene
PI	Proportional-integral
PID	Proportional-integral-derivative
PP	Polypropylene
PR	Peng–Robinson equation of state
ppm	Parts per million
ppmv	Parts per million by volume
ppmw	Parts per million by weight
psi	Pounds per square inch
psia	Pounds per square inch absolute
psig	Pounds per square inch gauge
PTB	Pounds of salt per thousand barrels of clean crude oil
PTV	Prover true volume
PTT	Petroleum Authority of Thailand
PVC	Polyvinyl chloride
RK	Redlich–Kwong equation of state
RP	Recommended practice (e.g., API RP 14 E)

rpm	Revolutions per minute
RVP	Reid vapor pressure
S	Sulfur
SAW	Submerged arc welded
S&W	Sediment and water
SCADA	Supervisory control and data acquisition
Scf	Standard cubic foot; means of expressing volume of natural and other gases. The volume at 60 °F and 14.696 psia (ideal gas) for process calculations. For sales purposes, it may be defined differently by law in some states in the United States
Scfm	Standard cubic feet per minute
SDV	Shut-down valve
SDWA	Safe Drinking Water Act
SF	Shrinkage factor
SG	Specific gravity
SI	Abbreviation for (1) shut in, (2) Système International (French for "International System of Units")
SP	Set point
SPE	Society of Petroleum Engineers
SRB	Sulfate-reducing bacteria
stbo	Stock-tank barrels of crude oil
TAPS	Trans-Alaska Pipeline System
TBP	True boiling point
TEG	Triethylene glycol
TVP	True vapor or bubble-point pressure
TTEG	Tetra Ethylene Glycol
UMSRK	Usdin-McAuliffe form of the SRK equation of state
UIC	Underground injection control
UOP K	Universal Oil Products *K* factor
USGS	United States Geological Survey
VLE	Vapor–liquid equilibrium
VRU	Vapor recovery unit
WC	Water column (e.g., h_w=80 in. WC)
WMT	Waste-management technology
WOR	Water–oil ratio
∇	Increment or difference

Index

Note: Page numbers with 'f' indicate figures and 't' indicate tables.